METALWORKING

SINK or SWIM

Metalworking

Sink or Swim

Tips and Tricks for Machinists, Welders, and Fabricators

Tom Lipton

Industrial Press
New York

Library of Congress Cataloging-in-Publication Data

Lipton, Tom.
  Metalworking sink or swim: tips and tricks for machinists, welders, and fabricators/Tom Lipton.
      p.   cm.
  ISBN 978-0-8311-3362-7 (softcover)
  1. Metal-work.   I. Title.
  TS205.L57 2008
  671–dc22

                                                          2008027549

Industrial Press, Inc.
32 Haviland Street, Unit 2C,
South Norwalk, Connecticut 06854

Sponsoring Editor: John Carleo
Developmental Editor: Robert Weinstein
Interior Text and Cover Design: Janet Romano
Printed in India by Nutech Print Services

10   9   8   7   6

ACKNOWLEDGEMENTS

This book might never have existed had it not been for a few important people. I would like to take a moment here to thank them properly.

Without my wife, Sargamo, I don't think this book would have ever been finished. She was able to cheer me on at key points and breathe some life back into me. As a fellow metalworker, she can read this material and understand it fully. We met in a welding shop 25 years ago and, for some unknown reason, she has not cut me loose yet. We have the dubious and unique honor of having the worst fight of our marriage over a pair of really nice C-clamps at the flea market. She can bring home the bacon as well as weld me under the table with one hand tied behind her back. In addition to her metalworking skills, she doggedly tried to improve my grammar and punctuation in micrometer-like steps.

I would like to thank all the metalworkers who have gone before me, on whose shoulders I am standing. Looking ahead to the future, I don't like what I see for the skilled trades. I am doing everything I can to make sure nothing dies that shouldn't. I have learned from so many people it would take a book of its own to thank everyone properly.

Chris Owen may not be what I would call a career metalworker, but I still owe him a little credit. He had the dubious honor of letting the book-in-progress out of the bag at WESTEC 2006 to an unexpected warm reception. By the way, Chris, you still owe me a lunch.

I believe that my parents had a strong hand in shaping my life as a metalworker—from the day when I was nine in the basement learning how to weld with my dad to my mom fronting the money for my first welding machine. How could I fail with support like that?

Thanks to all!

Tom Lipton
August 2008

TABLE OF CONTENTS

Diving In

Welcome to Sink or Swim

I have always wanted to write a book. I have started more than my fair share actually, but for numerous mostly lame reasons have never been able to finish any. I guess we'll wait and see how this one turns out. Let me know what you think. Given the small prodding of even a tepid response, I might have another one up my sleeve.

Career Metalworker is probably the best way to describe me. As I write this manuscript, forty-seven years have passed under my bridge since I was hatched and I still love metalwork. I may not have loved all the jobs I have done or the places I have worked, but I have always loved the trade. My parents might have had a small clue I was destined for the trades after my mom gave me the serrated saw off an aluminum foil box, which I put to good use sawing up the arms of her nice dining room chairs.

Maybe a more accurate description of me would be that I love the skilled trades. I can appreciate and be humbled by the violinmaker, the plumber, and the tugboat captain. The act of building something is deeply satisfying and difficult to explain to people outside the trades. Wherever humans pick up tools and work with materials, machines, and a skilled hand, this is where I want to be.

The only caveat is that it must be done well. This appreciation for attention to detail was drummed into my head by some of the old guys, more by osmosis and boot in the rear than any direct action on my part. I always felt that I let them down if I did a bad job or something didn't quite come out the way they wanted it. Most of the

Figure 1-1: A Steel Fabrication Shop at Full Throttle.

time, however, they didn't even have to say anything. You knew from the look on their faces or a callused hand sliding over the offending detail that you had somehow failed slightly. The answer for me was to try harder the next time and learn from the experience.

Call me strange, but I love the sight, sounds, and smells of a working shop (Figure 1-1). Each has its own distinct flavor and heartbeat. The smell of hot metal and cutting oil brings certain memories out in clear relief. Almost any welder can smell a piece of paper burning halfway across the shop in the middle of cutting steel plate with a torch, sniffing the air like a bloodhound looking for the start of a fire. We can tell which shop the boss sent the grinding work out to by the smell of their cutting oil. These shops we work in get into our blood in more ways than you know. The squeal of a tortured cutting tool, the clank of a pair of

1

vise-grips opening, or the sound of a tack weld breaking is as recognizable as your own name called by your mother in a noisy room.

My first experiences in metal working started with welding. My father taught me how to stick weld when I was nine years old in the dark basement of our house in Berkeley. Like the sailors who get the open ocean into their blood, I can say this is the moment I was infected by a fascination with metalworking or, at least at that point, welding.

In school I took machine shop and welding and never looked back. Somebody I thought was smart told me back then that having two different sets of skills was a valuable asset. They could not have been more right and it has served me well for a long time. The real message was never stop learning about your trade.

Young people just entering the trade are encouraged to stick with it and get through the tough beginning years. Things will still need to be built from metals and the trade needs new talent to advance. Be versatile and don't shy away from the tough jobs — you will be rewarded with a lifetime of support and hopefully enjoyment of a job well done. These first few years are the character building years where you "Pay your Dues" and learn an appreciation for all the aspects of your trade.

Personal Learning Attitude

Your attitude is one of the key ingredients to success in any field, not just metalworking. Without a positive and persistent attitude, you might as well just go sit in front of the TV and bathe yourself in some nice cable programming. The power of learning and dogged persistence cannot be overstated. My wife and I call it burning rod. You have to burn rod and put your time in to learn how to weld or become skilled in any trade. In my experience, most people don't learn on the first rod they burn or the first thread they cut.

Winners do what losers are unwilling to do.

We are in the middle of a unique time in history. The ability to share new ideas, information, and old skills will never be better. This critical time balances between the new guard and the old. On one side, we have access to technology for sharing huge amounts of detailed information across thousands of miles and time zones in the blink of an eye. On the other side, we still have access to the people and knowledge whose shoulders we are standing on and who form the foundations of our trades. This combination of factors has not been the case throughout history.

There was a time not that long ago where people never traveled more than ten miles from the town where they were born. Ten miles represents the distance you could walk and return home in one day. Anything outside that radius might as well have been imaginary.

Right now in our time, I can move my finger and click two or three times and look at the surface of another planet in our solar system. That to me is truly amazing. Now you can learn or teach all the way around the planet. Borders and time zones have no real meaning now for the learning process. You will either be in this wave of learning, or be left behind and fossilized by it.

This book is about learning new, and advancing current metalworking skills. The trades have been very good to me. Part of the requirement the trade imposes is to pass on knowledge and skills to those willing to learn. We have all stood on the shoulders of the people we have learned from; we owe at least the payment of passing the skills on. Each generation should push the boundaries of their art to the next higher levels. I thank the people I have learned from because without them I would still be trying to figure out how to smelt iron.

Your attitude toward learning and your skills are your protection in modern times. No longer can you rely on having a good job for life working for a stable company. Entire industries are being created or becoming obsolete on a daily basis. Modern skilled tradesmen have to constantly adapt and add skills to their toolkit to keep up with the pace of industry and the modern global electronic economy. Your skills must not stay static. Learn everything you can about everything. Sink, swim, or get the heck out of the way.

The advancement of any craft depends on new experiences and new people with sometimes wild and exaggerated ideas who push the boundaries of current knowledge or accepted practice. This is one of the character traits that built America. For this reason, as new ideas, techniques, and materials become available, it is important to postpone or suspend judgment about them. Look at how they might be applied instead of dismissing them. An open environment where every person and every new idea has worth — without concern for criticism or dismissal — is key to success. Truth and accuracy in knowledge and information, and the destruction of myth and misinformation, are required to further the art. Speak the truth and walk a reasonably toleranced line.

It never ceases to amaze me how some people memorize sports trivia or batting averages, yet more often than not they are the same people who ask how to run a particular machine in the shop they have been walking past for ten years. Instead of investing some of their time and effort to improve themselves, they choose to invest in a big screen TV or an F-350 4 × 4 turbo diesel to haul groceries. These are some of the same folks who will ask to borrow your tape measure because they don't have one.

These same people are typically the ones who never have enough of anything. More money, more beer, more toys, more horsepower. They don't correlate that their skills = value = profit = wages. Notice wages comes after profit. The companies we work for or start and run ourselves must be profitable or they cease to exist. Part of our responsibility as skilled tradesmen is to use our skills to make sure the companies we work for and start ourselves survive and prosper.

Every company is built on people. Machines and materials are commodities that can be bought, sold, and traded any day of the week. Great people are grown, cultivated, protected, and nurtured. In exchange for this, they give back loyalty, dedica-

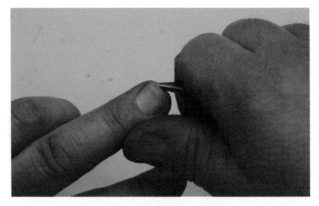

Figure 1-2: Meathook Hands.

tion, innovation, and hard work. It's called a trade for a reason. Don't misunderstand what I am saying here. I enjoy leisure time as much as the next guy, but I also love my work and would be doing the same thing even if somebody didn't conveniently pay me to do it.

You cannot learn skilled trades by reading a book, even this book, period. You can get an understanding of the technical issues and the tools involved, but true skill comes from hands-on practice. Anybody can learn some metalworking trivia and talk a good technical line. Just like a good salesman, they can sell themselves like a shiny new car. But, there is nowhere to hide out in the shop when the rubber hits the road. You either go up in smoke or gain traction. An imposter stands out like a cow-pie at a croquet match to someone skilled in the trade.

An interesting example of this comes from a story a friend told me about passing through international customs one time. The customs inspector asked him what kind of work he did, to which he replied that he was a machinist and worked with metal. The next question from the customs inspector was, "Let me see your hands" This is pretty telling, that you can judge a person's validity by looking at their hands (see Figure 1-2). All I can say is 'good luck' if you're an imposter!

Skills are like calluses; the faster you try to go, the easier it is to get a blister. The slow, steady approach builds skills and calluses for a lifetime of learning and rewards. If you think you can read a book or take a pill and miraculously emerge a

Figure 1-3: Charlie Blessing and Doug Duane. Master Toolmaker and Sheetmetal Man.

seasoned veteran, you are mistaken. It takes years to hardwire the necessary muscle memory to perform some of these operations, but once you have it is obvious to others in the trade. It's the little things that give away the masters — like the way a sheetmetal man flips his wrist to fan a cushion of air between sheets of metal or the little quick head nod of an experienced welder putting their hood down. You can't learn these things overnight.

I know for myself, and I'm pretty sure it's the same for most folks, that the way I learned is in little bits and pieces — gems and nuggets, if you will, during the process of making lots of mistakes. Slowly you gather these small parts together of the really big puzzle. That why old geezers are so darned smart; they have been picking up pieces of the puzzle for a long time. These veterans have had a lot time to gather and polish their nuggets. The trick here is to get them written down or passed on

before your memory starts to fail. That's what I'm trying to do with this collection of nuggets.

Somehow I have been lucky to develop good relationships with several great teachers (see Figures 1-3 and 1-4). I sometimes feel pathetic and puny next to their skills. The only way I could ever hope to surpass them is if they die and give me a chance to catch up a little.

That's exactly what's happening.

My old teachers and workmates are dying off. When they die, they become static points. All their amassed skills stop growing and start to dissipate until they disappear forever or, worse, have to be learned again. By writing down and documenting as many as I can remember, I can preserve them for future tradesmen. So do your part for the trade. Take some old geezers under your wing and learn something from them. The baby boomers will be retiring in droves in the next several years, taking all their wonderful hard-earned knowledge with them right out the door. I have tools that were given to me from some of the most influential people I learned from. Every time I pick up one of these tools and use it, a flood of memories comes back along with a deep appreciation and feeling of humbleness. I can almost hear them telling me to be careful and not screw up that nice tool I gave you.

In ten thousand years, modern humans will have forgotten how to read the ancient language this book is written in. But I am willing to bet they will still use metals and need to fabricate them into useful articles. I'm sure the methods will be different. However, I am confident they will have their roots in the things we know now and have learned from the people that went before us.

Shop Environment

All work and no play make for a pretty dull shop. Working in a shop with a bunch of other people is somewhat like a marriage. There are good, bad, and really funny days. Just like a family, there are members you get along with and others that you don't. You spend more waking time at work with your

Figure 1-4: Fred Van Bebber. Master Machinist.

workmates than you do with your mate or immediate family in a given work week. If you can't have a little fun and get along, it makes for a pretty miserable time. I have purposely included an attempt at humor in some of the descriptions and pictures.

Over the years I have worked in many shops, some large and some small. Overall I prefer the small shop dynamic. The flavor of a shop is created by the people working there. Shops can foster and nurture the learning and skill building attitude or they can undermine and destroy it. It is a choice.

One of the greatest gifts I have been given is the thirst for skill growth. This sounds simple enough, but is much harder to do in practice. If you were to ask anybody if they support skill growth or learning, what do you think they would say? In all likelihood they will agree and say "yes." The only way to truly judge this is by actions. How do you support this by action? Humans learn best by doing things. In particular, things they are interested in. When you are interested in something, the learning is almost effortless. Here are two scenarios to think about.

You're sitting in a classroom listening to the teacher talk about how long it takes a train leaving Chicago to get to New York if it's traveling east at 60 mph and another train leaves New York, yada, yada, ... This is what I call linear or structured learning, otherwise known as lacking moisture and inductive of sleep. Almost all schools and academic institutions use this method. They start at the beginning and move in a deliberate step-by-step fashion until they get to the end. Some people thrive in this type of cranial learning environment. Some of these folks end up as engineers or scientists along with business managers. Normally, they are politely called white-collar workers by shop folks. These are the same people who write textbooks and decide how to train and improve the other type of workers. You know, the ones with blue collars and dirty jeans.

Most trades people have learned their skills in a much different way. In fact, some may have gone into the trades because they didn't like the structured linear method. The main method of learning skilled trades is the direct hands-on method. In my experi-

ence, the learning bounces around more and is definitely less linear than in a school environment. Typically whatever you were working on was the subject of the lesson that day. Most of us blue-necks have come to be where we are by this bouncing around method. Have you ever been surprised by how different and incomplete the descriptions of tasks and operations you read about in a book were when you actually tried them out for real? If you were lucky enough to learn some of your craft in an apprenticeship program, then you know what I mean. This balances the need for some theoretical work with a healthy dose of doing things in the shop.

Thursday Nights

Fairly early in my career, I was very lucky to work at a shop that supported hands on skill growth and learning in a unique way. I don't think they realized what they were doing and it certainly was not intended for the purpose of training. I only realized it many years later when I was in the position to implement a similar setup.

The shop allowed us to work on small personal projects using the company facilities and resources.

It sounds pretty dumb and simple, but there is quite a bit here. One of the old guys I worked with at this company was employee number 001. He was the first employee and for many years he was the shop foreman until a serious industrial accident sidelined him. He started the tradition of what I call Thursday nights. This was the special time set aside for guys to work on their own projects. It was only one day a week but it was sanctioned, albeit weakly, by the company. We were allowed to use company equipment and minor materials and build almost whatever we wanted. This special time was set aside so the inevitable "G-Jobs," or personal projects, would not be done on company time. He used to say,

"Every man has a little bracket in their life."

Thursday nights had a much more important effect.

What happened was the guys would build things. And more important, they built things that they were interested in. They were gain hands-on experience in the best possible way, by doing. You could build almost anything including things you had never built before. Things the shop foreman would never assign to you because you didn't have the skill or experience. Obviously there were some limits on what you could put together. The most extreme example I know of took place over a fourteen-year period when a friend built a forty-eight foot sailboat. He built the boat in his backyard but almost all the fittings and bits crossed his workbench at one time or another. Another guy built a stainless steel hot tub. The list goes on and on.

Well, I say, if you have the experience, who cares how you came by it? So by doing things and trying things you had never done on Thursday night on your own time, you built up your skills. I look back and the most successful people at that company were the ones that were there every Thursday night chunking away on their own projects and "brackets." Most success, either career or financial, can be traced to proactive learning behavior.

Here are the basic rules we use to this day.

- Two or more people minimum working together. No one works alone. Somebody has the duty and responsibility for locking up.
- You cannot disturb any company work in progress. That is, unless you finish it. I have seen quite a few company jobs completed on

personal time just so a machine could be used for a minute.
- You cannot run a business out of your employer's shop. That means no work for money. Trades and barter are okay, but no visitors. The only exception to the no visitors rule is if they bring food.
- You ask permission to use any company materials and pay for any major materials. The foreman has the say so on the amounts and types.
- You do not use company floor space to store personal works in progress. It goes home every night.
- You clean up your tracks. Ideally the process should be invisible. Nobody will complain if a six-pack of sodas shows up in the lunch room fridge once in a while. Show your appreciation by example, not talk.

Figures 1-5, 1-6, 1-7, and 1-8 show a few of the things I have built over the years with my

Figure 1-6: Helve Hammer.

Figure 1-5: Thursday Night Personal Projects.

Figure 1-7: Pyramid Rolls.

Figure 1-8: English Wheel.

Thursday night program just for the love of building things.

This kind of program provides the fastest, most effective way I know of to gain important skills. I support this program in the shop I manage, with excellent results. The program also delivers a super-positive message from the company to the employees, supporting them by trust and positive action. The company invests the materials and machinery along with space and the students invest the time. The rewards are the students gaining skill and position while the company gains skills, versatility, loyalty, and dedication. What more could you ask for in a fair trade?

Of course, you can have a classroom and books along with written tests. But the true measure is: can they do the work? "They passed the written test but they can't find the start switch" is the all too common result of book learning. How many drivers' licenses are handed out by just passing the written test and a minimal hands-on demonstration of driving skills? I see the results of that method every day on the highway during the sleepy morning migration and the angry afternoon free for all.

You really have to push the pedals and get dirty to learn this work. You can't learn skydiving or how to pull nine G's in a fighter plane without getting off the ground. Another example of this idea shows up clearly in a hiring situation. Suppose you have your choice between two candidates — one who had practical experience and time in the shop doing a particular operation, and another who had taken a class and passed a test but had not been in the shop. Which candidate would you favor?

What's a Journeyman Anyway?

There are many names for the seasoned accomplished tradesman. I didn't make up the names. Typically throughout history, these jobs and professions were dominated by men. The names reflect that fact. It is not intended to be derogatory or sexist — only historically accurate.

Journeyman, Apprentice, Tradesman, Master, Rookie, and Craftsman are all names associated with different skill levels related to the skilled trades. Most trades have no published standards or colored martial arts belts given out to indicate specifically what it means to be an apprentice or the tenth-degree grand master.

In my mind, it is not necessarily the number of years served in a particular trade but more a question of ability. All too often people are given a title just because they have a certain number of years at the bench. I have seen 40-year veterans who stopped learning after their second year and became miserable static points in the trade. Yet I have also seen 4-year apprentices, who were literally sponges starved for information, easily exceeding their static counterparts. There lies the problem. How do you measure ability? The definition of ability is different for each trade and cannot be measured merely by the passage of time. The only answer is for other top people in the specific trade to establish and make the ability assessments.

The French apprenticeship association (Les Compagnons du Devoir du Tour de France) has a system of skill measurement that seems perfect. It has withstood the test of time, 400 years and counting. After a certain number of years in a particular trade, you must submit a project for your masterwork. No term paper, no book report or thesis, but some real down-and-dirty work. I guess they figure if you hang around for four or five years, you will have at least learned which end of the hammer goes down, but they still want you to prove it.

This project proposal is reviewed by a panel of masters in that trade. If the project is judged difficult enough to demonstrate a high level of skill and competency, then you're off and running. No special time off or preferred treatment is allowed. The project must be completed along with all the other responsibilities the student has. Gee, it's kind of like the real world — pressure included.

The completed masterwork is presented and judged by a panel of masters in the particular trade. If it passes scrutiny, the applicant is awarded their master card (pun intended). In the case of the Les Compagnons, it is a cane or staff with the colors of their chosen trade on it. If the project is not deemed difficult enough or will not demonstrate the proper combination of skills, it is rejected. The applicant must then submit a new project or modify the original project. If you don't complete the project, you get to stay mucking about in the lower levels forever. If students do a lousy job, then I imagine they have to wait to try again.

My definition of the top meat eaters of the skilled trade food chain goes like this. It's the people who have enough skill and experience to draw on in order to take on any problem that comes up in their trade. They may not know exactly how to tackle every job, but they have the experience and acuity to chip away using their skills and experiences to get almost any job done well. Journeyman cavemen can catch, cook, and clean their dinner as well as make a more efficient spear from one of the leg bones — and then go out and do it again and again day in and day out.

With all that said let's get to the good part. Good luck and never stop learning.

"The price of failure is only knowledge."

Format

Finding the best format has been one of the hardest parts of writing this book and has contributed to the failure of previous efforts. How could I present this kind of information to readers in a way that keeps them interested, yet is not so bloated and long winded that the true gems of information become lost. In the end, I decided to use a format that reinforces the way this kind of information gets into our heads in the first place.

Call it a recipe book with each item standing on its own like a good meal and becoming a part of the whole experience. Lots of pictures contribute to a magazine-like format that can be snacked on in little bites any time you feel like opening the book.

Not all the ideas in this book will be of use right away. Your mind and situation have to open and ready to receive. Others bits will be immediately useful whereas some may never be. Heck, I'm pretty sure I will even be accused of "copying" ideas. Remember folks, the knowledge belongs to the trade, not to any individual, me included. I will be the first one to admit that I learned many of these things from other people and by keeping my ears open and my mouth shut.

Figure 1-9: Damn, I think I left my tape measure at the other end of the shop!

I hope this book will be something to refer back to and even add recipes of your own to become a kind of larger metalworking cookbook. Many things written here are directly related to my own personal experiences. That does not automatically make them right for everybody. Use common sense and decide for yourself what makes the most sense to you.

If you are offended by anything I write here for whatever reason, please try to get a grip on reality; otherwise you will never survive in a real shop. In fact, let me say that anybody who is offended by this material is a prime candidate for a life of torture and harassment in pretty much any metalworking shop. Don't let the informal style of my writing fool you. Trust me; there are some great nuggets in here. Potentially even one of the suggestions you find here could repay the investment in this book a thousand times over. I admit this material doesn't read like a Tom Clancy techno-thriller, but if you enjoy your trade, you should find some useful information in here and I almost guarantee at least a chuckle or two.

There are many excellent, well-written books on the basics of these subjects. Some very good ones are listed in the bibliography and recommended reading list. This book is designed intentionally for metalworkers who already have a solid background in the basics, like righty tighty lefty loosey. I make the assumption that the readers already know whether their rear ends have been drilled, punched, bored, or reamed. . . .

Brain Food

Not all the skills we need in the metalworking shop involve the hands. Our brains are the most powerful tools we have at our disposal if we will just pry them open a little. Many times we can think our way out of a problem if we give ourselves half a chance. Improvement in thinking starts with admitting you don't know something. Have you ever had a serious discussion with someone you admired and thought was extremely intelligent? If you pay attention, when you talk to people like that you will notice that you end up explaining many things in great detail about your areas of expertise to them. Why is that? I think it's because for years they have been able to admit they don't understand something, which opened their door to learning. They are able to ask even the most mundane questions because they readily admit they don't know something. As soon as you say "I don't understand," you open the door. You can't put money inside a locked safe; the door has to be opened and the combination starts with "I don't understand."

Communication

Communication is accomplished in the metalworking shop in several ways. The more effective you are with these methods, the further you will go in your chosen trade. Effective communication is quick, clean, accurate, and — most important — to the point. A good engineering drawing has all these elements. Have you ever noticed that the best drawings have no questions? Effective written and visual communication can be considered a form of language. If you don't speak the language, you are condemned to get your information from people who do. A stranger in a strange land is not in a friendly place.

A short note or sketch is worth its weight in gold compared to even the most detailed verbal information (Figures 2-1 and 2-2). If you just counted the parts in a box, take a second and leave a note for the next person so they don't have to re-count (Figure 2-3). The note is a physical representation of a single thought that all you have to do is see it and you instantly remember.

When you have to turn over a job to someone else, leave a brief note, even if it's just to say the job is finished. Some of these things are plain old manners that your mother should have taught you already.

Figure 2-1: A simple note prevents confusion.

11

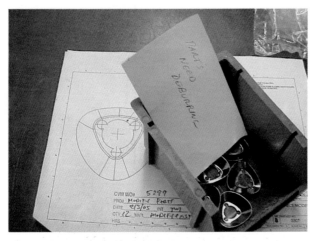

Figure 2-2: Leave a note for the next person.

Figure 2-4: A physical template.

If you must receive information in oral form, do yourself a huge favor and write it down immediately. A few quickly taken notes may just make the difference in doing a job correctly. Documentation wins over memory when the chips are down and hats are being handed out for rear ends.

A simple elevation of your trade would be just to learn how to spell, or at least use words that you can spell correctly. I'm still working on this one. . . .

Make accurate physical templates and take copious written notes in the field (Figures 2-4 and 2-5). Don't alter the information that a template made in the field is communicating. The template made in the field and the written note are gold when you're back at the bench. Typically it is a real pain to get a second chance to take field measurements.

Therefore, when you do have the chance, overdo it. Pretend a lawyer will be looking at your templates and field notes and asking those kinds of questions only a lawyer can ask.

An inexpensive digital camera can collect information faster than an army of draftsmen or court stenographers. These are so inexpensive and easy to use, there is no excuse why every shop doesn't have one. Take an extra few shots from slightly different angles. You never know which photo will become the real lifesaver. My rule is when I think I have enough pictures, I take ten more for good measure. It's the pits when you get back to find out some photos are out of focus or the flash washed out the critical detail you needed. This problem became painfully obvious when I started this book project. Digital pictures are basically free data. Most of the time they are never printed but can provide invaluable documentation.

Mark or otherwise identify bad parts as soon as it is known they are bad (Figure 2-6). They may still have some other use, but the parts must be

Figure 2-3: Let others know what has been completed.

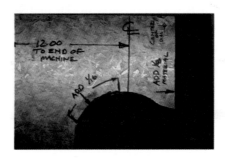

Figure 2-5: Notes on a template communicate important information.

Figure 2-6: Identify bad parts.

Figure 2-8: You don't need paper to write a reminder!

unmistakable and easily identifiable from the good ones. This labeling communicates to others without interpretation or question what the status of a part is. If you don't mark them immediately, then others who pick up the part will have to figure it out themselves, driving the cost of the scrap part even higher.

Do your roughing with the most foolproof measuring tool you have (Figure 2-7). As you refine your dimensions, change to the appropriate tools for the level of accuracy required: scale, calipers, and then micrometers. It's hard but not impossible to screw up a 12-inch dimension when using a 12-inch ruler to take the measurement.

Machinists don't make mistakes in one-thousandth increments. They miss by .100 or 1.00. The classic mistake is one revolution of the micrometer barrel, or .025. I've seen some gut wrenching mistakes made because of one turn of the barrel.

Welders and fabricators typically miss in one-inch increments. When you make your first measurement, be sure to force yourself to read the actual number, then the fractional part. You would be surprised how often this mistake happens.

Leave a note for yourself if you need to remember something (Figure 2-8). I sometimes write on my hand in a pinch. It's easy to forget something if you have a lot of things to remember, so write as many down as is practical. Another trick I have used if it is really important to remember something is to leave my car keys with whatever it is I'm supposed to remember. You won't get very far without realizing that you are supposed to remember something. Mark your drops and scraps early (Figure 2-9). If something is worth saving, you had best know what it is.

Ask questions early and make sure you understand the answers. There is nothing worse than missing something and having to ask again and again. It's really easy to say "I don't understand" or "I'm not with you" early on. It's much harder to ask later after you were supposed to know.

Figure 2-7: Roughing with a reliable measuring tool.

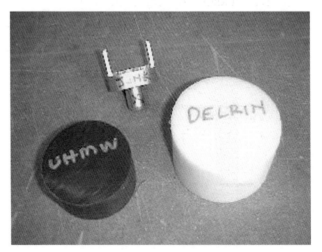

Figure 2-9: Labeling saves time later.

Drawing and Sketching

Drawing is the common language of our metal-working professions. Except for minor differences in conventions, two metalworkers in different countries should be able to exchange drawings and be able to interpret the specifications. This ability to communicate across verbal language barriers is extremely powerful. We can drive a car in a foreign country because of the common factors related to driving. Drawing and sketching are the same. Making up your own drawing conventions and rules of the road is just like really bad driving — everybody hates it.

Use 8-1/2 × 11 sketch paper. A bad sketch on small paper is a worse sketch. Those little free pads the metal suppliers give you are meant to make you screw up your work with bad small drawings so that you have to buy even more metal. . . .

Draw large and fill up the full extents of the sheet. Tiny sketches are like tiny brains: who wants one?

You can't borrow a sheet of paper. Have you ever had one returned to you?

Press hard with your writing instrument. Faint lines and text are for accountants, not metalworkers. I like to see the little blivits where the lead actually broke because you pressed so hard. Everything shows up when it goes through the copier if you use

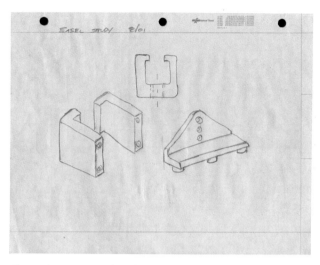

Figure 2-11: Example of the clarity of shop floor isometric hand sketches.

a heavy hand. Anybody who uses a pencil with a .5mm lead does not press hard enough.

Take some time and learn how to functionally sketch in the isometric projection (Figures 2-10 and 2-11). This technique saves countless hours of drawing time because many times you only need one view instead of the standard two or three views to fully communicate a part (Figures 2-12 and 2-13).

I have a little trick I use to help me with my isometric drawing. I made a special ruler which, combined with normal graph paper or even plain lined paper, can make everybody's isometric drawings look pretty good. You can buy ruled isometric paper at the drafting store or from some office supply places. The problem I have found with the

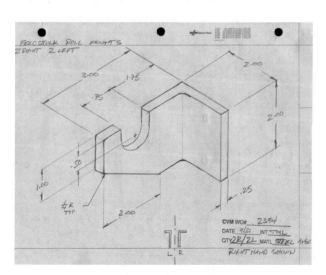

Figure 2-10: Isometric projections.

Figure 2-12: Another isometric hand sketch.

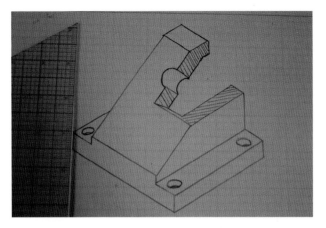

Figure 2-13: Another isometric hand sketch.

Figure 2-15: Using an ISO ruler.

pre-printed isometric paper is this: when you photocopy or scan it, the grid lines get darker and wash out your sketch object lines, making your nice isometric sketch a jumble of spidery lines.

If you use some common sense and a few guidelines, you can easily make decent looking isometric sketches right on the shop floor. You will need a plastic ruler that's clear so you can see through it (Figure 2-14). It should have lines on it parallel to the long axis of the ruler. These are used to set distances off parallel to existing lines. They also work great for spacing hatch lines when you draw a section view.

If you want really nice circles you will need an isometric circle template. I skip the template out in the shop and hand draw these. Always remember

that the goal is to get it done as quickly as possible with good quality. It's always better to have a decent quick sketch than a perfect sketch that took way too long to make.

You can see the Iso-ruler I made is cut off at the same angle as the normal isometric axis, approximately 35 degrees in this case. The see-through plastic and lines are important for setting off parallel distances (Figure 2-15).

The three axes seen in Figure 2-16 are the isometric axes. When drawing to scale, these are the only directions that measurements can be made along. With the Iso-ruler you can align the angled edge with the vertical gridlines of normal graph paper or even the edge of the pad itself. Flip the angled edge over and you have an alignment guide for the opposite direction. I like the engineering pads that have the lithographed grid on them for anything I will run through the copier. The lines are a faint green and disappear in the copier. These pads are made by National and distributed by the Avery office supply company.

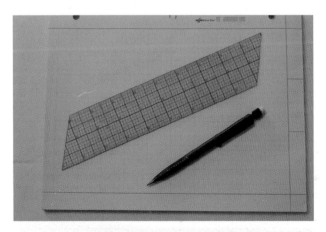

Figure 2-14: See through the ISO ruler for cross hatching and setting off parallel lines.

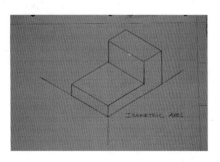

Figure 2-16: The three isometric axes.

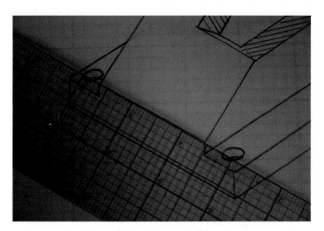

Figure 2-17: Using see-through grid to set parallel lines.

With the parallel lines on the ruler, you can set off distances quickly by sighting through the clear ruler (Figure 2-17). This is a great simple method for drawing quick, concise shop sketches. Just remember: it's better to have some kind of sketch even, if it's not technically perfect. You just can't do this stuff from pure memory.

When quick sketching, try to get the scale and proportions to look right. They don't have to be deadly accurate; they just need to look about right. If a part is long and skinny, try to sketch it that way.

Learn and endeavor to use the basic rules of drafting.

Think you know what the basic rules of drafting are? Open a drafting book and open your mind. These rules and conventions have been refined over hundreds of years. You can bend them a little, but don't break them into a million pieces.

Some of the major abuses I see in modern electronic drafting are incorrect line weights, omitting hidden lines in views, missing dimensions, incorrect third angle projections, lazy details on hole bottoms, and thread misrepresentations. The list goes on and on. These abuses result from ignorance of the conventions of drafting and plain old laziness. Just because you can draw something with an expensive electronic drawing program does not mean you are doing it according to established conventions. As the saying goes, garbage in, garbage out.

Avoid the temptation to use an electronic drafting or modeling program right out of the gate on a new design. In my opinion, in the earliest stages of a design, CAD is often a hindrance — or worse, a waste of time. CAD promotes a microscopic view with way too much detail and precision in the opening stages of a concept development. This weakness, coupled with the possibilities of countless revisions and iterations, can make CAD a serious time liability. If you can't get the concept down with paper and pencil, you have no business behind the mouse of a computer.

Keep your original hand sketches in a binder. Having these handy has saved me many times. Always release a copy to the shop and retain the original information as backup.

As a subject, geometric tolerancing, otherwise known as ANSI Y14.5, is a whole can of worms on its own. Entire books and endless dead dog beating discussions have been devoted to the subject of ANSI Y14.5 tolerancing. I agree with the overall idea and philosophy, but the execution has been less than perfect. Hundreds of drawings cross my desk every year and I see every misuse of this system in the book. Interpretation and confusion caused by a system that was designed to help eliminate these same issues has wasted countless millions of man hours. It literally has turned into the Swiss army knife of tolerancing systems. Use it carefully and with deliberate thought and purpose. It's kind of like a loaded gun in a crowded elevator; you really want to be careful where you point that thing.

Here are some of my own personalized favorites. I call it ANSI WISEGUY 14.55 (see Figure 2-18).

Hey, if we can poke a little fun at ourselves, what's the point? I would like to have rubber stamps made and mark up the drawings that misuse the symbols and send them back to where they came from.

You can convert old sketches into new sketches. Many parts are boringly similar. Use this similarity to your advantage when sketching. I have a couple of hand sketches without the dimensions — I copy them over and over to save sketching time

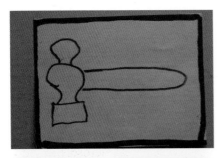

Figure 2-18(a): Press Fit.

Figure 2-18(b): Deadly Feature.

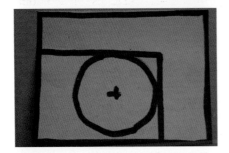

Figure 2-18(c): Radius Too Small.

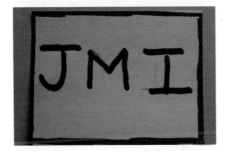

Figure 2-18(d): Just Make It.

Figure 2-18(e): Day Shift Feature.

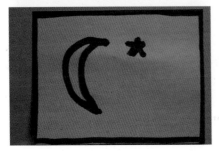

Figure 2-18(f): Night Shift Feature.

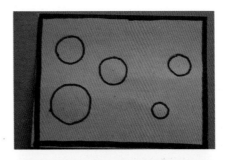

Figure 2-18(g): Extra Holes OK.

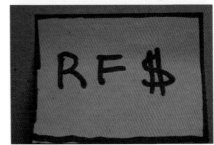

Figure 2-18(h): Regardless of Feature Cost.

Figure 2-18(i): Close.

Figure 2-18(j): Really Close.

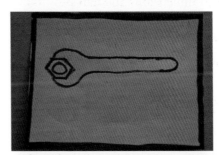

Figure 2-18(k): Tight.

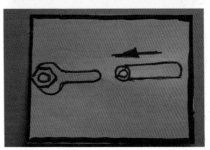

Figure 2-18(l): Really Tight.

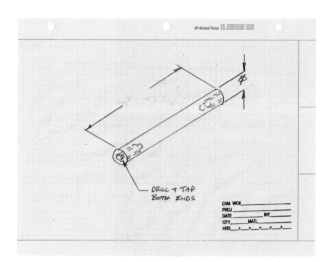

Figure 2-19: Converting an old sketch.

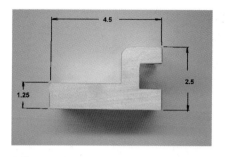

Figure 2-21: A digital picture of a part.

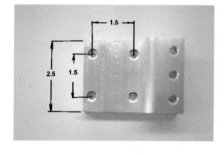

Figure 2-22: Add dimensions to a digital picture of a part.

(see Figures 2-19 and 2-20). Just fill in the blanks and, whammo!, a new drawing of the same old boring part.

You can also print out a digital picture of a part and dimension the photo to create a decent drawing (Figures 2-21 and 2-22). Sometimes you can lay a part directly on the copier or scanner and get a decent print that way.

Laser printers are so good now that you can print to scale and measure the lines with calipers to double check yourself or find a missing dimension (Figure 2-23). If the designer used a computer, there is a better than 50/50 chance the drawing is correctly drawn to dimension. You can use this to

your advantage to determine a missing dimension even if the paper drawing has not been plotted to any particular scale. If you measure a dimensioned distance, you can calculate the scale of the plot and derive missing dimension pretty accurately. Obviously it's better to have the real numbers, but some times you have to get a part out regardless of any missing information.

To save layout time, use light contact cement to bond a full-scale laser drawing directly on your material (Figure 2-24). This is a great prototyping trick. Put tiny circles at the centers of

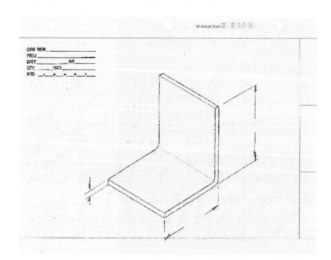

Figure 2-20: Reusing old sketches can save time.

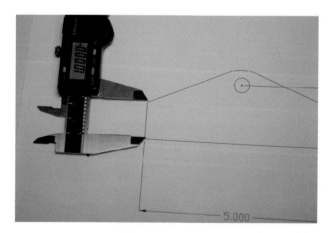

Figure 2-23: Using calipers to find missing dimension from an accurate laser plot.

Figure 2-24: Bonding a full-scale laser drawing.

any holes you want to make. A .015-diameter center hole printed out makes a great center punch target.

You can make a larger 1:1 full-scale drawing with an 8.5 × 11 printer by making a mosaic (Figure 2-25). You can even have little alignment marks to make the mosaic assembly easier. A long, skinny full-scale drawing is a good example of something you can use this trick on. Figure 2-25 shows an assembly printed out 1:1 that would not normally fit on 8.5 × 11 paper.

Figure 2-26 shows the approximately 8.5 × 11 print sizes overlaid on the drawing. There is really no limit to how large a part or assembly you can tile together using this method. For some jobs having a 1:1 scale, drawing is pretty handy.

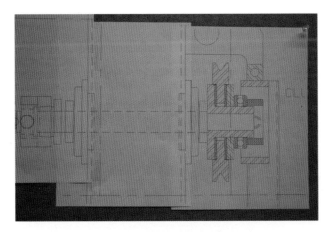

Figure 2-25: A mosaic drawing made up from several smaller full-scale drawings.

Date and initial all your hand sketches. It improves communication and documentation, and everyone will know who is making those great sketches.

A quick five-minute sketch before a rush job has saved many a metalworker from painting themselves into a corner (Figures 2-27 and 2-28).

Use whatever is handy to communicate in a written form. I call this method Table-Cad. I've done some of my best design work right on the table. Just don't spill the acetone.

Minimizing Screw-Ups

Some of the best learning and a good portion of skill development come from making mistakes. Why is it when the job goes smoothly you can't remember how you did it? But when you struggle or make a huge mistake, the memory is much more imbedded and clear. Treat your mistakes as learning tools. It is even more important to know what not to do than to think you know.

There are a million ways to screw up in this business. I invent new ways almost every day. Learn from other people's mistakes as well as your own. Planning and anticipation of potential problems are hallmarks of people who have made enough mistakes to know better.

Think the job all the way through. A few minutes making a plan will pay you back tenfold. Change your plan if it looks bad. Plans should be flexible and dynamic to account for unexpected events.

Admit your mistakes quickly and move on. Dwelling and denial waste time and valuable energy. Mistakes are common so get used to them. The difference among mistakes is in how much times goes by before they are discovered. The trick is to catch the mistake before much time has passed. Check frequently and consistently.

It's common for a metalworker to make mistakes on a simple job but have no mistakes on a complicated demanding job. The difference is attention. Pay attention on the easy ones.

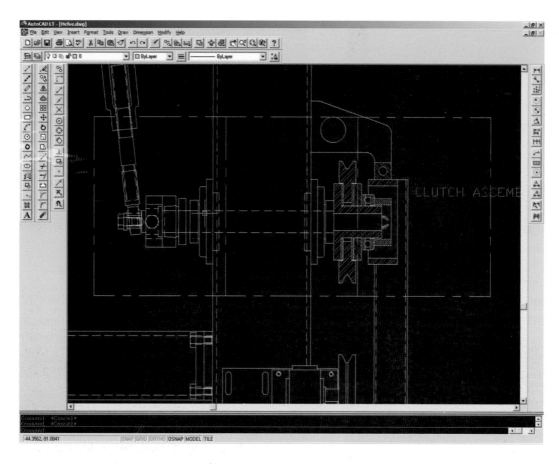

Figure 2-26: Tiling together a large drawing.

When in doubt opt for maximum material condition. Doing so may give you the option of going back and removing more if you need to.

Checking your work often with attention can minimize the time a mistake can exist.

Watch out for the right hand left hand or mirror scenario. Many times to save drawing time a mirror image is called out in the notes or work order. Don't fall for this one.

The classic dunce mistake is to miss the quantity. It's the pits when you make one part and three were called out.

When in serious doubt or confusion, go back to something you absolutely know is correct either by

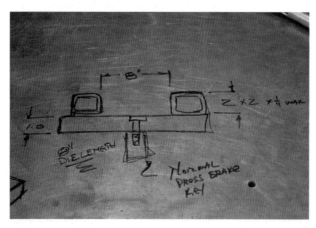

Figure 2-27: TABLECAD.

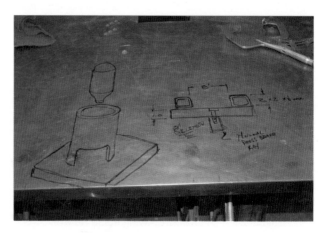

Figure 2-28: Quick sketches save time during rush jobs.

Figure 2-29: Avoiding catastrophic errors.

direct measurement or observation. Work your way forward from this point. It's dangerous to make any assumptions when in doubt.

When tracking down errors, make no assumptions. Verify everything when doing a forensic investigation. That means you take and read the measurements yourself. Don't just assume that the other guy did his part correctly. Verify it.

When you can't find a direct cause, look at cumulative errors and effects. Small amounts here and there can add up and throw something significantly off farther down the line.

There are some classes of work where errors can be catastrophic (Figures 2-29 and 2-30). Have you noticed that these jobs always go to certain people in the shop? Watch these people carefully and learn their methods. If you find yourself in this

Figure 2-30: Some jobs are restricted to selected workers.

position, slow down and think the job all the way through. If you can do this kind of work, everyone will leave you alone — guaranteed.

If all else fails and you still make a mistake, here are a few options you may not have thought of.

- Be sure to hide your scrap parts in the middle of the scrap barrel. If you hide them in the bottom of the barrel, then when the barrel gets dumped out, your dirty underwear is on the top for everyone to see.
- And another old standby is to blame the guy who just got fired or quit.
- Paint it black and ship it at night.

Somebody once told me a funny story about a shop that was on the waterfront somewhere. The shop had been there for quite some time. The machine shop bay doors opened onto a narrow estuary separated by a wooden dock that ran along the side of the building. One day there was an unusually low minus tide. To the shock and horror of the foreman, littering the exposed mud of the estuary were hundreds of scrap parts. Most of the stuff was piled about the distance you could heave something from just inside the bay door. Apparently when something got screwed up it got tossed into the water hopefully never to be seen again. Whoops!

Accuracy

Some ways of becoming an accurate metalworker are related to the habits developed in the quest to minimize mistakes. Attention to small details along with deliberate and frequent checking help.

Real accuracy comes from your work habits and skills, not the particular measuring tools.

Don't let the dimensional accuracy get out of control before you take action. Small, measured corrections are always better than gross corrective applications.

You cannot expect to last long if high accuracy is your only goal. Tolerances are stated to allow for material variations and manufacturing efficiency. Learn to use them. You cannot justify producing half the work at twice the needed quality and tolerance

Figure 2-31: A tool and a nickname given to a very fast and dear friend of mine.

requirements just because you can while your work mate churns out lower quality parts that nevertheless meet specification and are within tolerances. In other words, don't build a bridge with a micrometer when a ruler will suffice.

Speed

One trait that contributes to speed is decisive decision making. Sometimes it's more important to make a decision than to try to make the perfect decision.

If you realize your decision was wrong, make a new one and proceed again. Hand wringing and whining are a brain drain. Allow for mistakes and variables in your plan. Don't find yourself in enemy territory and out of bullets.

Take the long view. Try to anticipate things that will be needed and when they are needed. Don't let the lack of a simple part or tool stall your entire job.

Identify critical aspects of the job early on and have a contingency plan if something changes or doesn't go as planned.

Lead from the front. Demonstrate by confident example. These are the hallmarks of a great teacher and an even better foreman. A leader is also the first guy to get shot in a charge out of the trenches. It takes a certain kind of personality trait that some people may lack.

Keep your momentum up. It's easier to keep the job moving with small consistent energy inputs. If it stalls, a much larger input is necessary to get it moving again. Inertia does work.

So now your brains hurt and you want to look at more pretty pictures of interesting tips and tricks, fine with me. There will be a test on this chapter Monday morning. . . . .

Shop Math

Here are a few tricks that can make you look like a genius with half a brain or at least half a head (Figure 2-32).

This is another area where a little practice goes a long way. I have found over the years that many metalworkers became metalworkers because they hated math in school. To me this seems weird because of how much math there is in the metalworking trade. If you want to get ahead and be one of the top dogs in the shop with your own bed and bowl, you will need some math in your toolkit. This section is not meant to be a complete course in shop math — only a few useful nuggets to make you at least look like you know what you're doing. I didn't invent any of this stuff; some old guys who had way too much time on their hands thought of it years ago.

I was the same way as many people when I first started out. I didn't like math and avoided it like you might avoid an IRS tax audit. My turnaround

Figure 2-32: Math tricks can be helpful in many ways.

came when I found myself in a tight spot — I was tapped to teach a welding class basic shop math! I tried my best to weasel out of the trap but there was no escape; everybody else had already out-weaseled me. The weekend before the class started, I hit the books pretty hard trying to get a leg up on my students.

What I discovered during that painful period was all I had to do was go slow and stay a little ahead of everybody else. I learned two major life-time nuggets teaching that first class. First was that a little practice is all it takes. A strong motivator, like having to chew your leg off to get out of trap, helps a lot also. And the second nugget was the more you practice the better you get. It sounds lame, but it's very true. Teaching that class several times firmly cemented the knowledge. It also gave me the confidence to go much farther.

I have found that skill in math is one of the major dividing lines between the trades and the sciences. It's more like a stumbling block, or the tar pit that the dinosaur gets stuck in. This one skill has limited more tradesmen than I care to estimate. The worst part is there is no reason for it other than lack of trying.

In reality, there are a small number of concepts and formulas you actually need to memorize. The real trick is to leverage what you have memorized to analyze other problems. Almost all math problems can be solved in several distinctly different ways. Sheetmetal layout is a great example. You can use trigonometry to figure out the compound angles and true lengths found in many sheetmetal layouts. Or you can use the graphic methods that are taught to all apprentice sheetmetal workers. The results are equal in all respects. In fact, the graphic method was taught for the specific reason to avoid mathematics, which was the perceived realm of educated scientists and engineers — some kind of throwback to forbidden knowledge.

Let's talk about the circle first. This is one of the most important shapes we will ever encounter in the trades. It pays to know a few of the facts related to the circle.

Figure 2-33: Drawing a circle.

But first a little story about why you really, *really* want to learn about the circle (Figure 2-33).

The Slurpee Program

At one place I worked, two brothers also worked in the shop at the bench next to mine. They were hired as kind of mid-level mechanics who did a little bit of everything including minor seagull welding and rough machine work.

Now I know it takes all kinds to make the world go around, but these two guys were a piece of work. I never really figured out how they lasted as long as they did. On one level, I was glad that I didn't have to do some of the miserable jobs they were given.

I think between the two of them they might have had two years experience in welding and half that in machine work. They were actually pretty good at some of the real quick and dirty kind of jobs that always seem to be around a metalworking shop.

Fairly often they would ask me questions about how to make different things or how to set up one machine or another to do what they wanted. If you're like me, you go ahead and give them the answers even when you know they should have figured this out themselves by now or remembered it from the last time they had to do it.

One day the older brother came up to me and asked me how to figure the length of a sheetmetal blank needed to make a cylinder of a specific diameter. Now in the sheetmetal world you learn how to do this right after you learn how to use the potty all by yourself. To make matters worse, I had told him at least two other times exactly how to do it. I even wrote it on his workbench in sharpie the last time. I guess he spilled the acetone on his bench because he pleaded with me to tell him.

I guess this was the straw that broke the camel's back, so to speak. I blasted him up one side and down the other. At the end, he actually was begging me to just tell him how to do it so he could get his job done. I caved and told him one more time with the caveat that I would never tell him again. He either had to write it down or ask someone else. He thanked me profusely and off he went.

When I came back from lunch later that day, I found a supersized 7-11 Slurpee waiting in a little puddle of ice cold condensate on my workbench. It was the middle of summer, so it was a welcome surprise even though the shop was air conditioned.

After a bit, I saw the older brother and asked him if he was the one who dropped the drink off. He nodded and thanked me again for helping him. I don't know about you, but I find it sure is nice to be acknowledged for doing somebody a favor. It feels downright good. I thanked him for the tasty beverage and thought to myself that if people acted like that more often, it would make it a lot easier to be nice to people. Then, suddenly, I snapped out of that daydream.

I don't quite remember how long it was until the younger brother came up and asked me how to make something he didn't have a clue on how to start and even less of a chance of completing. It must be a very difficult position when you have been assigned something that you know is simple, but you don't know how to even start. And to top it off, you will have to ask somebody who's going to bust your butt before they give you the answer.

About halfway through the blasting, I realized my lips were a bit parched. Offhandedly, I suggested that if he brought me a tasty frozen beverage back from lunch, I would open Pandora's box of knowledge and tell him all the secrets of his latest problem. I was surprised at how quickly he agreed.

Well I'll give you two guesses what was waiting for me on my workbench when I came back from lunch. And the first guess doesn't count.

I had just invented the Slurpee program.

For a while it was great. I even got to the point where I knew what flavors were available on each day of the week. Out of sheer boredom, I invented complicated mixtures of half grape one quarter lemon lime and the rest whatever red stuff they had, just to try and throw them off. I can tell you these guys had a lot of questions and once they had a system for getting the answers they used it. In actual practice, if you implement a similar system I could suggest a few changes to make it even better. You might try the "Steak Sandwich program" Or the "Fill my gas tank program" instead.

I retired the Slurpee program when I realized I had gained ten pounds and my tongue had a semi-permanent purple tinge to it.

The moral of the story is this: learn everything you can to be self-sufficient or be humiliated into paying for information.

Every metalworker should know the terminology and properties of a circle. Figure 2-34 shows the basic parts that you should understand without question. These come up so often in our field during the course of everyday problem solving that without the knowledge of these properties we are handicapped.

I have tried to keep some of the more useful properties of the circle within easy reach in my severely-limited cranial memory bank.

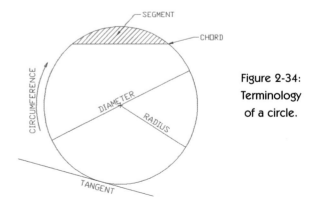

Figure 2-34: Terminology of a circle.

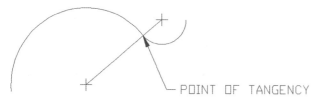

Figure 2-35: The line connecting the centers of the two arcs passes through the point of tangency.

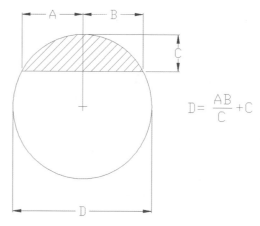

$$D = \frac{AB}{C} + C$$

Figure 2-36: Measuring the chords of a circle.

A line tangent to any point touching a circle is perpendicular to the center of circle at the point of tangency (Figure 2-35). This fact is useful when doing inspection and reverse engineering.

Two tangent arcs of different radii will have their centers along a common line passing through the point of tangency. (Figure 2-35)

The arc length of any circular arc can be quickly calculated by multiplying the radius by the arc angle by .01745. In formula, $r\alpha(.01745) =$ Arc length. You will see this number again; it's one of the ones worth remembering.

Hopefully we all can remember that the circumference of a circle is the diameter multiplied by the ratio π. If not, mail me a Slurpee and I'll tell you the answer. . . .

True roundness or circularity cannot be measured with a two-point measuring system An example we have all seen of this is with poor, center-less ground shafts and pins. These will measure correctly with a two-point measuring tool like a micrometer. But when these pins or shafts are spun in a v-block and compared with an indicator, they show some roundness deviation as runout. The only true way to measure roundness or circularity is by rotation of an accurate spindle or center point.

I have found the formulas relating to measuring chords to be extremely useful (Figure 2-36). I have often used these calculations to find an unknown radius. Many times things in the field or in the shop have a radius that is outside the range of standard radius gages.

You use a bar or pin of a known length to act as the chord for your measurement. Finding the radius can then be reduced to a simple calculation. The space between the pin and the radius to be measured can be determined with gage pins or by measuring off the top of the pin with calipers (Figure 2-37). Be sure to measure in the center of the pin or bar to get the maximum measurement to the radius. A depth micrometer can be used for this purpose also. The tip diameter of the depth micrometer will affect the depth measurement, but you can add a ball tip to your measuring tool to get around that problem or to make some additional correction calculations for the tip diameter.

Another slightly different formula for chords is, $R = \frac{C^2 + 4H^2}{8H}$. This formula is arranged to give the radius of the arc instead of the diameter (Figure 2-38).

Figure 2-37: Using gage pins and chords to measure an unknown radius.

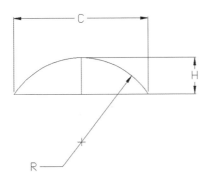

Figure 2-38: The radius of an arc.

Mass, Volume, and Area

Another math topic that always seems to crop up is the measurement of weights, volumes, and areas. The typical questions go something like this.

How much does that plate weigh? How much would it weigh if we switched to aluminum instead of steel? How many gallons does that tank hold? Or, perhaps, how much does that tank weigh when it's full?

If you keep a few common facts in mind, you can handle these problems on the fly when they come up. The real trick is to try to remember only a few key items that you can keep in your head and leverage to derive other things. I'm pretty happy if I can figure something out in my head and be within 10% of the correct answer. A few that have seemed to stick in my head and be useful over and over again are:

- There are 231 cubic inches in one gallon.
- Water weighs 8.33 lbs per gallon. Or the shorthand method, "A pint is a pound the world around," which gives the approximation of 8 lbs per gallon.
- 1 pound per square inch water pressure is a column 2.31 feet tall. This is easy to remember because the numbers are the same as the number of cubic inches in one gallon.
- 1-inch-thick steel plate weighs 40.8 lbs per square foot. I tend to round off the decimal and just use 40 lbs per square foot. This gives an answer accurate to 2%.
- The density of steel is approximately .283 lb/cu". You can round this off to .3 and be within 6%.

- Aluminum is approximately 1/3 the density of steel. Most aluminum has a density pretty close to .10 lb/cu", depending on the alloy. This is easier to remember than the exact number. It's also an easy factor to divide or multiply by. If you want to know the exact numbers, 6061 is .098 per cu in., 2024 is .101 per cu in, and the rest are so close it's almost irrelevant.
- The United States five-cent coin (the nickel) weighs almost exactly five grams. It is so close, it's scary.

Angles and Shop Trigonometry

I've mentioned before in this book that angle work can be some of the most difficult work facing a shop worker. Laying out and measuring angles on the vast number of projects that a job shop might encounter requires a good grounding in trigonometry and geometry. You can get these from a large number of other books; our focus is how these things relate to our work in the shop actually building things.

Here are a few of my personal favorites that I use over and over again.

Sohcahtoa

The way I learned basic trigonometry concepts was through the story of the old Indian chief Sohcahtoa. If you can remember this story, you can easily solve any right triangle problem you might bump into during the course of your work.

Here is how it breaks down if you have not seen this before.

SOH = Sine θ = Opposite/Hypotenuse
CAH = Cosine θ = Adjacent/Hypotenuse
TOA = Tangent θ = Opposite/Adjacent

That's it. Memorize this and you will have right triangles nailed. Another thing some people have not seen is to get a direct angle result use the shift key and appropriate trigonometric function for your problem, which on most pocket calculators raises the function to the power of negative 1. This one extra step will display the actual angle instead of the trigonometric function of that angle. The key will look like this — TAN^{-1} or SIN^{-1} or COS^{-1}. Applying this step saves you a trip to the trig tables, at least if you didn't already know this.

Square Root of 2

Another number I have found to be extremely useful is 1.414. The ratio 1.414:1 is valuable when finding angles and lengths related to 45°. In fact, 1.414 is the square root of 2. The 45° angle pops up in the shop on a daily basis. Remembering 1.414 is very useful for figuring chamfers, spot drill diameters, and the cut lengths of gussets, diagonals, and braces.

Here is an interesting real world example of the relative sizes of angular divisions. When I used to shoot at the target range, I learned to appreciate just how small some circular divisions are. The adjustments on telescopic sights for rifles normally move in 1/2-minute divisions with some going as small as 1/4 or even 1/8 minute clicks or adjustments. Now at 100-yards distance or the length of a football field for our non-shooting audience, 1 degree of angle is 60 inches (Figure 2-39). A good shot with a decent rifle can shoot under one minute of angle which is 1/60 of a degree or one inch at 100 yards. You probably already know that one degree contains 60 minutes of angle and each minute contains another 60 seconds of angle. They actually have an infinite number of divisions, but these are the commonly-accepted conventional divisions of the circle.

Our one inch corresponds to 1 minute of angle at 100 yards, so 1/60 of one inch is 1 second of angle at 100 yards distance. In turn, 1/64 of an inch is pretty close to 1/60 of an inch, so approximately 1/64 of an inch is 1 second of angle at 100 yards. I have 64ths divisions on my pocket scale that I carry around everywhere; I hardly use them because they're hard to read at book-reading distance, let alone 100 yards. That's a pretty small angle. The Moore Precision Tool® company makes an eight-inch rotary table accurate to 1/10 of an arc second — yikes! That about .002 inch at 100 yards!

Sine of One Degree

This little trick is handy for mentally calculating angles and rises when you don't have a calculator handy. Its basis is the sine of 1 degree, more specifically that the sine of 1 degree at one inch length is .0175 (Figure 2-40).

This ratio of .0175 per inch scales quite well. If you increase the angle to two degrees, you can double the .0175 to get .035. All you need to remember to apply this handy trick is .0175 is the rise of a one-degree angle at one inch in length. If you double the length, then double the rise. If you increase the angle by five, then increase the rise by five.

It scales well up to quite a few degrees of angle. I generally use it only for estimating small angles under 10 degrees.

What is the rise of an angle of two degrees at 12 inches? Using this method we would double the sine of one degree (.0175) to get .035 and then multiply .035 by 12 to get .420. If you ask the old Indian chief Sohcahtoa and apply some simple trigonometry, you get .419, which agrees closely. Try it yourself with a few angles to assure yourself of the scaling. I have found this to be a very useful

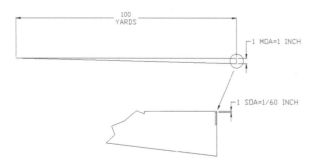

Figure 2-39: Angular divisions.

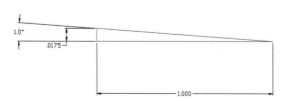

Figure 2-40: Graphic illustration of the relationship of the Sine of 1 degree.

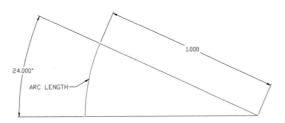

Figure 2-41: Calculating arc lengths.

technique. A typical question you might want to answer using this trick is: what is ±1 degree at a 10-inch radius? .0175 × 10 = ±.175

You can expand on this technique to calculate arc lengths:

$$(r) \times (\theta) \times .01745 = arc\ length$$

For small angles, the rise and the arc length are almost the same. This technique works best for calculating longer arc lengths where the angle is much larger. For example, in Figure 2-41, the arc length is

$$(1.0)(24)(.01745) = .419$$

The Metric System

Can we please have a little practical talk about the metric system? I know it is supposed to be better and soooo much easier to use. My main gripe with it is being caught in the middle of a change that will realistically take the United States another forty years or longer. Some industries are extremely reluctant to make the switch because of the huge investments in infrastructure dedicated to imperial measurements. How can we be expected to change when the basic infrastructure is so deeply reluctant and lacks any incentive to comply. Until I can get materials, drawings, and supplies readily in metric sizes, I will keep one leg firmly planted in the English world, thank you very much. If we are forced into compliance, it will only cost us — with no return on the massive investment to change. The folks working outside of the required compliance net will reap the benefits of our added overhead; they will make parts and pieces as good, or better, than we do at lower cost.

Problems come up in engineering and metalworking when designers and engineers are switching between the different systems. I call these "hybrid" drawings; the effects manifest themselves in the little details that always get messed up anyway like tolerances and surface finishes. The alarm bell goes off when you see a really tight tolerance on an otherwise mundane metric drawing. There is a huge difference between .005 inch and .005 mm. We so often get badly drawn and badly checked drawings that adding the further burden of switching measuring systems seems to fall under the category of wants instead of needs. What I really want is good drawings.

In reality, the metric system holds no particular advantage to the machine tool industry. I agree that, in some of the sciences and true engineering, the metric system makes some sense because they are actually manipulating the numbers to obtain a solution and unit conversions are simplified. In the machine tool industry, we merely read whatever numbers we are given and make the parts accordingly.

Think about this a little.

The machines read the numbers they are fed and have little meaning to the machine programmer or operator. Who looks at 50,000 lines of numbers with any comprehension? As long as we stick to decimals instead of fractions, we are using a base ten system which looks suspiciously like the metric system. In fact, you cannot tell the difference. Is the gripe with Imperial really only related to fractions? No, the real problem is units. The basic stumbling block has always been and always will be units. Hey, we put guys on the moon with slide rules and the English system — sounds pretty good to me.

"Hey Neil, how far are you from the lunar module? Over."

"Ahh, Houston, I'm three decimeters from the module. Over."

Are metric measurements easier to record? Not at all. They are just numbers that we read from our tools and compare to the numbers on the drawing. Is a M6-1.0 plug gage read more easily than a

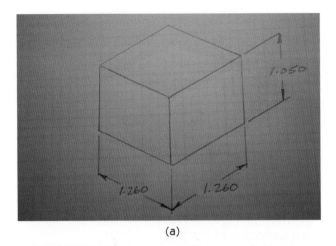

(a)

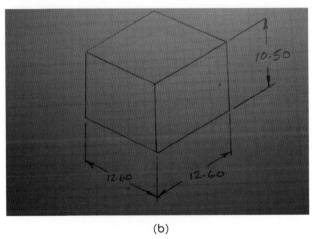

(b)

Figure 2-42: Which drawing is metric?

1/4-28 plug gage? Does the metric system help you with your trigonometry? Think about it this way. If the numbers on a drawing were replaced by letters or symbols, could you still make the part? Sure, as long as all the letters you needed were on the drawing and your tools were calibrated in those letters. So we find ourselves back at the need for *good drawings*, not necessarily a particular system of measurement. The metric system cannot possibly help with the problem of bad drawings.

In Figure 2-42, guess which drawing is metric. Everything is meaningless — including the metric system — unless we specify the units.

To illustrate how confused everybody is, I saw a kid on TV being asked how big a particular dinosaur was. His reply was "Fifty meters long, and 130 feet high." He probably memorized these facts from two different dinosaur books

and went ahead like a good future engineer and mixed the units. That in itself is not a problem because the kid at least gave us the units he was using, which is better than most of the drawings that cross my desk with the dimensioning mode set to "free for all."

Another friend of mine who is an aspiring metalworker told me that when he needs to take really "accurate" measurements, he switches to millimeters. I think because the lines are closer together on his plastic ruler he stole from the kindergarten kid down the street, it just feels more accurate to him. I don't have the heart to tell him how those pesky little 100ths are even harder to read. I think he also has a pair of those scissors they give you in fifth grade with the rounded ends to go with his thermoplastic ruler.

Another one of my favorite metric stories is about a job a friend did in his shop for one of the big government labs. They gave him a nice detailed metric drawing to manufacture a part. All the dimensions were on the drawing so no questions ever came up. When the part was delivered, the "scientist" asked why the part was so small. Well it turns out he was using centimeters to draw the part and it was manufactured using millimeters. Whoops, Houston we have a problem, the satellite just burned up in orbit. . . .

I don't really love or hate the metric system. I just don't like having the metric system rammed down my throat by somebody telling me it's easier to use or we won't do business with you unless you are "metrified." I guess I'm what you could call "pro-choice" when it comes to the metric system.

So the next time somebody tells you how much easier the metric system is, quickly ask them how tall they are or what their body temperature is. If they reply in feet and inches or degrees Fahrenheit, you can then send them away to calculate their height in some metric units. At the same time, ask why, if the metric system is so easy, they didn't just use it. Or if they reply with metric units and ask you what your height or temperature is, reply in some unexpected metric units like nanometers or degrees Kelvin and see what kind of reaction you get. Hey they're part of the metric system — just convert it if it's so easy.

Computers and the Metalworker

Every day computers gain a little more foothold in almost every industry in the world. There is no doubt that computers have changed the face of the planet forever. We might argue about whether the change is for the good. Many metalworkers deliberately shy away from computers and the machinery that is controlled by them. Why is that? To the modern metalworker, the computer is just another tool that allows them to do a different class of work or the same class in a different way.

My observations have led me to believe that, in most cases, computers are an age barrier. There seems to be an almost clear dividing line in ages: on one side you have computer acceptance and on the other side you have primarily distain or outright rejection. This is not a concrete rule by any means. There are many crossovers from both age groups.

I read a little story about technology that went something like this.

Before the invention of electric refrigeration, there were professional ice cutters. In the winter, they would go out onto the frozen lakes and saw blocks of ice from the lake for use during the rest of the year. The ice was stored in insulated barns and caves. At the time, the craft had evolved into a mature and efficient industry. They had designed special lifting and transporting equipment to efficiently cut lift and move the blocks for the best labor economy. During the remaining part of the year the ice cutters delivered and sold blocks of ice to households and businesses to keep their food and products cool. The term icebox is related to this time period. They readily accepted technology improvements that improved existing operations — up to a point.

Then along comes some clever person who discovers that using expanding gases and a compressor can produce cold and even freezing temperatures without the need for ice. The ice cutting industry wholly rejected the technology, partly because it directly threatened their livelihood. The technology was new and radically different.

When is the last time you met an ice cutter?

Computers are like refrigerators in this story. They are new and complicated. They directly threaten many people's livelihoods. I'm also pretty sure they're here to stay. As with any new skill, the first part is the most painful. I can say with a straight face that learning how to use a computer is a lot less painful than some of the other skills I have learned over the years.

As a metalworker, you need to learn a minimum of three skills with the computer or go the way of the Wisconsin ice cutters. Start by thinking of it like a tool in your toolbox.

Learn how to use a drafting program on the computer

If you decide you want to learn only one thing, this would be it. So many other skills related to computers a metalworker will encounter start with this one. It doesn't matter which exact program you learn, only that you learn one well enough to function with it and make a good drawing. Once you have learned one, the transfer to another is a thousand times easier. You already have a major head start in electronic drafting because you know what a good drawing looks like and the names of the different elements found in an engineering drawing. And by deciding to learn electronic drafting, you are already way ahead of the kid whose last job was making sandwiches.

Keep in mind the microscopic effect that computer design fosters. Computers cause you to focus on too much detail early in the design process. In the early stages, a pencil and paper are the right tools. If you are scratching your head with a pencil on how to do something, you should think twice about jumping on the computer.

Learn to make, take, send, receive, and edit digital pictures and files though e-mail

Every day more and more of the information we need to do our jobs comes to us through little copper wires or optical cable. In the modern world, if you don't have an e-mail address or a home computer, you have your picture in the dictionary under "Metalworkersaurus". If you don't have a computer, all I can say is every day you don't will make it that much harder to catch up and put you one step closer to fossilization.

A picture is worth a thousand words. I think everybody has heard this one before. So if you have a picture *and* a thousand words, what's that worth? Even the ability to ship drawings around the world in one day pales in comparison to the speed of electrons inside copper wires or optical cables. In the modern world, one day is the difference between getting the job done and missing out completely.

Do yourself a huge favor — get a digital camera and let the learning begin. You will need to learn not only how to take the pictures, but also how to do minor editing and re-sizing. This skills will only make your experience that much better. You can share and exchange vast amounts of information around the globe using the Internet and your new camera. You might meet a Danish sheetmetal worker whom you introduce to your New Zealander panel beater friend. Or you might develop a friendship with another inquisitive tradesperson looking for the same answers as you from eight time zones away. The connections are limitless. A handful of years ago you had a very slim chance indeed of meeting any of these people, let alone learning something from them or interacting on any level. I have personally traded parts for a lathe for a pail of pickled herring with a nice person living on an island in a fjord in Norway through the power of the Internet. Thanks Ole!

Learn how to search for information and resources on the internet.

Every day I am amazed at the information that is available to us. I am also appalled at the amount of garbage and useless chaff that circulates with it. Fortunately, as humans we can still filter the unwanted swarf better than any computer or the software that runs on it. It admittedly takes some practice to use these search tools to their highest good.

You will get no argument from me that computers can be seriously frustrating at times. I don't love computers I just want to use them and put them away like any other tool. I also expect them to work when I need them to work — but that subject is a whole can of worms on its own. The only other advice I can offer is to stick with it and try for the steady learning of little bits and pieces. Here is some other general advice to computer users.

- Weigh the efficiency a computer can add or subtract to a task. It is not always the correct decision to use a computer for every job.
- Sometimes it's harder to see the bigger picture on the computer screen. At some point something will have to get made in the shop. A quick mockup in the shop can save you many burned-out eyeball cells in the office.
- Many welders and machinist are so fast they can actually make something in the shop for testing before it can be drawn up on the computer. Don't underestimate these guys and their abilities. You can try three ideas out in the shop before you can detail one completely on the computer. Sometimes there is no substitute for a physical prototype. It has a presence like no drawing ever has.
- Beware of the microscope effect computers have. If you're not diligent, you can find yourself stuck down in the details when all you wanted to do was check a dimension. Don't miss the whole forest by burning the ants into cinders with a magnifying glass.
- Be realistic about the information the computer gives you. It may calculate internally with dozens of decimal places but that in itself does not assure accuracy. The computer has no idea of the existence of the real world and the imperfect humans who inhabit it.

So hopefully you saw or read a couple of things in this chapter you have never seen or even thought about and can put them to good use.

Dumb and Dumber

This is a little steam venting section were we can all have a little laugh at some of my pet peeves. I'm sure we all have some special ones of our own, but this is my book and I get to rant a little and possibly plant an idea of two in a few folk's heads. Don't worry — this is a short section with some funny stuff in it.

We as humans are inundated with ridiculous systems, products, and people on a daily basis. There is so much dumb stuff going on that it's easy to become numb to the dumbness. Intense advertising and marketing has stuffed our bellies full of crappy products and services until we are so bloated we cannot even decide for ourselves if something is well designed or engineered.

Design and engineering are among those things that can be really great changing your life for the better or can be criminally lame. Advertising and marketing blur the line between the two to make them indistinguishable from one another. As consumers, we are often not included in all the considerations that are weighed when a product is being designed and developed. From the looks of some things, it seems the legal department is taking on the bulk of new product design. I often wonder how many interesting products never get to market because of this worrying about lawsuits. This process of profit by litigation is weakening us as a country known for innovation and "Wild West" risk taking.

Product design is such a subtle art. One or two bad decisions and you can make an otherwise good product bad or, worse, annoying. Well-designed products and tools are a pleasure to use and are immediately obvious in superiority. This goes double for superb personal service. I'm sure everyone has some of their own favorite examples. The smooth operation of a fine mechanical device and the silent mind-reading waiter in your favorite restaurant are the results of deliberate attention to detail and practice of your particular trade.

The pinnacle of Elmer Fudd engineering is realized when product designers attempt to recover from a complete lack of thought. They spend the bulk of their time wrapping their product in a bright finish that has the look of some fancy, organic, industrial design. Adding insult to injury, unnecessary complexity is often included just to be able to claim the award for most features.

Just make the thing work like it's supposed to please. A tool should look, act, and feel like a tool. Can we just agree on that? Maybe I'm not the best person to judge the finer points of esthetic design, but I sure as heck can judge good function. Remember: form follows function, not the other way around.

Here are some of my peeves that, if I were king, I would banish from the face of the earth.

Self-checkout lines

This idea almost offends me. I have a choice when and where I exert my economic power. If more people exercised their options when they make a purchase, then we would see rapid improvement in this area. When I walk into a business I am three-quarters of the way toward spending my money. Car salesmen are acutely aware of this fact. Why do you think they pounce on you like a cat on a small, blind, three-legged scurrying mammal as soon as you set foot on the car lot?

Figure 2-43: The joy of parking lots.

As I make my purchase selections, many times I have to make small compromises as I go. Now, finally, I am ready to use some of my hard-earned economic horsepower residing in my wallet and they want me to check myself out? Who the heck thought of this? I can just imagine a bunch of MBAs sitting around high-fiving each other when somebody suggested this one. I guess they figure once you're in the door they can go ahead and drop the ball completely as well as hide it. Most of these megastores can barely keep the shelves organized so you can find things. What makes you think they have fully considered the self-checkout system? We have to draw the line somewhere. If we let the corporate collective do it, I guarantee it will mean more work for us. If they really wanted to make things convenient, you could fill up your cart and just wheel right out the door without stopping, your account debited with 100% accuracy and zero waiting.

What's next, self-service parachute packing, or maybe do-it-yourself laser eye surgery? In our modern world, there is still room and need for specialists and expertise in all industries. One morning we might all wake up and discover to our horror that we don't know how to do anything except consume like beached whales or sell each other cheap junk made in another country that does remember how to make things. I'm sure we can all relate to what a pleasure it is to see a job well done no matter how simple it might seem to an outsider.

The scissor jack on my wife's minivan

I know these are designed to be manufactured as cheaply as possible and not intended for everyday use, but they are borderline next to being useless, scrap-barrel material. The handle has no ratchet and can be only swung through a pesky little arc before you have to flop it over to the opposite side. The screw is about as free running as a snow plow in a sand dune. To get the jack anywhere near the frame of the car takes about three hundred flip flops of the pathetic handle. All the while, you are crouching like a rodeo clown at a bull riding competition trying not to get killed on the side of the road. And then you think you're smart by leaving it extended when you put it away so the next time you only need to crank it a few turns. Well, the smarty-pants designers made the spot where the jack is stored fit the fully-collapsed jack like a latex glove that's two sizes too small. I guess I will be sure to renew my AAA membership, forever!

How many times have you seen these jacks tossed into the trunk or rear compartment because they are too much trouble to put back where they belong? Why can't these things be built into the car chassis? Push a button, or insert a handle into a hole in the body and off you go.

Metal paint can lids

Enough already. It's time to redesign this one. In the age of fancy blow-molded plastics, packaging designers could do wonders with this one. You shouldn't need two different tools to open and close a consumer product. Think bleach bottle, easy type packaging. Crusty lids and obliterated labels — be gone! Happily, I am seeing this exact gripe changing as I write this. Now I can complain about why it took so long to happen.

Parking lots

These are areas where all the rules of the road go out the window without any sidewalks. Pedestrians really need armored personnel carriers to make it out of the parking lot free-for-all and into the stores where they came to spend money. Most parking lots seem more like holding pens for cattle that are on the way to hamburger heaven. Parking lot designers should be condemned to wander through their creations for eternity blindfolded and naked pushing a shopping cart missing two diagonally-opposed wheels! All this fun while frustrated parkers encourage them with cattle prods.

How about safe walking zones where cars can't get you? A clever designer could even make these lanes automatically return the carts to the front of the store. Or better yet have your own personal shopping cart. It could fold up and attach to the back of your car. It would be yours to customize and equip to fit your lifestyle. Best of all you don't have to return it to the front of the store or dodge traffic to pretend to return it (instead, shoving it into an unoccupied parking space with one wheel barely hooked preventing it from rolling while you quickly escape).

Music CD packages

The little tab says "pull here." It should say "pull here to frustrate consumer with immediate seal failure and mandatory sticky goo application." I fail to understand why these packages have to be hermetically sealed to ten to the minus six torr. Is it just theft, or just a test to see if you have what it takes to outthink the packaging? Some folks might say it takes a little practice. Well, I say I don't have to practice opening the CD player to put the disc in to play it. Why should I have to "practice" opening the crummy ill-conceived packaging? The little tax labels on liquor bottles don't take any practice to open. I wonder why.

The lame cart that your welding machine comes with

I don't think the designers ever lifted a 330-cubic-foot argon cylinder up the twelve inches to get it over the lip of the cart and into the "convenient" bottle retention area. It's like they built the machine and then said, "Holy cow! We forgot to put a place to hold the gas cylinder. Quick! Get the duct tape and glue gun and add one on before we put it in the box." How about one that lowers to the floor so you can roll the bottle right into place?

People who insist on calling extruded bar stock "billet"

Look up billet in your metal supplier's handbook. Anybody who uses this term to describe anything other that a real billet should have a big red letter "D" stamped on their forehead for "dunce." This term is inaccurate and nearly offensive to anybody who knows better. How about machined from slab or bloom? Let's kill this one forever. In fact,

the next time somebody who knows better misuses this term in your presence, just slap them! I'll mail you a quarter.

Aircraft grade aluminum

What are people trying to say with this one? Airplanes have almost the full spectrum of aluminum alloys in them. Get a little more specific, folks. We are taking about a trade that is full of exact, precise, scientific, and engineering terms. This is purely a marketing ploy to free you of some of the weight in your wallet. Nobody says, "Sewage grade stainless." I wonder why.

Electrical plug strips

Why can't the designers take out their tape measures and measure the bulbous transformers that are mated on the ends of all modern electronics? It's bad enough they don't trust us with high voltage 110 anymore. The pitch between the electrical receptacles should be such that transformers can reside in peace with a modest side yard with their neighbor transformer. Or better yet, build the transformers internally with the plug strip and use a small standardized connector for the peripheral equipment. In reality, it's most likely a manufacturing expedient. If you have to send electrical products to several countries, it's probably easier to send a different voltage transformer instead of a complete high or low voltage product.

Modern handbooks that still have trig tables in them

Let's pull ourselves by the bookstraps into the electronic age. In the time of five-dollar pocket calculators, trig tables are just fluffy filler material, somewhat like packing peanuts. Please see Appendix 1.414 for the trig tables in this book. . . .

The slack-jawed teenager who makes submarine sandwiches

I bet you a dollar you can't order your sandwich with a single, clear concise sentence without incurring further inquiries from the sandwich maestro.

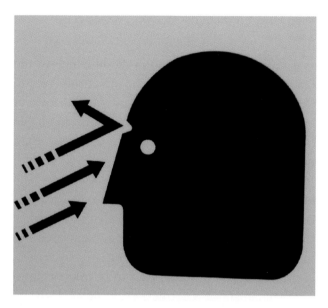

Figure 2-44: It's not that difficult!!!

The memory tree has not sent down roots yet. This country is in big trouble in the skilled trades department when the only hands-on jobs kids can get are fabricating multi-laminate meat-and-cheese energy capsules.

Want to Make a Million Dollars? Some Things That Really Need to be Invented

Here are some ideas out of my personal "Big Book of Million Dollar Ideas." Go for it if you like any of them. I don't have enough time left to work on a tenth of the things that need to be invented to make my life easier.

- Real time coolant refractometers. These would be plumbed into a machine's coolant system and give alerts when coolant concentrations are not within settable parameters. Put a knob on the front of the machine. Heck, the machine can control the entire thing for all I care.
- Built-in oil skimmers for CNC machinery. They all know about tramp oil contamination. Why don't most machines come with one already installed?
- Cheap easy memory upgrades for CNC

machines. Only recently have the machine tool builders addressed this with larger memory capacity as standard equipment. Gee, it took only thirty years to get it done. The prices should be on par with PC memory upgrades. This is a great opportunity for the computer geek crowd.

- Carbide inserts optimized for top performance and chip control on plastics. Ever try running your lathe unattended turning Nylon or UHMW? Didn't think so.
- Angular digital readouts for manual milling machine heads. These would also allow you to monitor the tram condition of the head at a glance.
- Right angle attachments for manual mills with a quill feature. Ever try tapping with the right angle head?
- A cheap direct pressure indicator for manual vises and chucks. This would give the operator some valuable feedback. Or at least help calibrate the apprentices.
- Computer controlled active chatter cancellation. Sympathetic frequencies injected into the spindle under precise computer control used to cancel chatter and squealing. Somewhat like active noise reduction headphones. Mechanical engineering students in need of a graduate study subject please apply here.
- A one-eighth-inch-diameter super-rigidium end mill that can cut four inches deep for all those designers who don't bother to think about how things get made. As soon as this is invented, some numbskull will need it to reach 4.2 deep.

Some other things I want to try

I always thought it would be interesting to completely switch careers once in a while. Not many people who have established a foothold in one trade get to branch off and try a entirely different career. I think there is much to learn about your own trade by examining the work and skills of other disciplines. I also think that this would be a two-way street, with both parties coming away with some new ideas to think about and push the boundaries out a bit.

One way I thought of doing this was to offer a week of vacation or 40 hours of free labor to learn something about another craft. Just hanging around with open eyes and mind, you would absorb a huge amount of information. Is it worth an investment of 40 hours of your life? I think it would be. Just trade a month's worth of lousy television programming for a new set of skills.

A few careers I wouldn't mind switching with:

Tugboat operator. Stone carver. Musical instrument maker. Wood boat builder. McMaster Carr order filler. Surveyor. Astronomer. Glass blower. Physicist. Industrial photographer. Biologist. Metal spinner. If anybody out there reading this willing to trade some teaching for free labor, drop me a note and let's see if we can work something out.

Shop ideas I want to try sometime when I have a minute

I thought a boring head in the tailstock of the lathe might make a good taper offsetting fixture. Or it could be used to align a worn tailstock to bring it on center. I have never tried this, but it has hung around in my head for quite a few years now.

I have never done any metal spinning. The process looks very interesting to me. I saw a picture once of the titanium shell for Jacques Cousteau's underwater saucer being spun red hot in a hydraulic metal-spinning setup.

Modify a few different files to fit in the clamping mechanism of a cordless reciprocating saw. If you found one with variable speed, this might make a nifty little addition to the toolbox.

Bean Counter Lounge

If you're an engineer or designer and you work in a place that has metalworking facilities under the same roof, you are a very lucky person. It is both rewarding and fun to see one of your designs come to life right before your eyes, especially if it works. There are some challenges inherent with working with shop personnel that they don't teach you in any school. I hope the reader will find a few useful ideas to make life in the shop a worthwhile and less painful experience.

In all my years as a metalworker, I have worked with dozens of engineers and scientists. From the metalworker's perspective, this experience can be very rewarding or pure bamboo-under-the-finger-nails torture. I can name on the fingers of two hands the engineers who earned my admiration and full respect. Not to say that all the rest were bad, just that the really good ones stand out in comparison. I nominate a special place in engineer Hell for the truly bad ones. This is the Hell where it's always cold and noisy and the only work they get to do is lay out the parking lot sprinkler diagram.

The hallmarks of these successful professionals were a combination of ability, empathy, and respect.

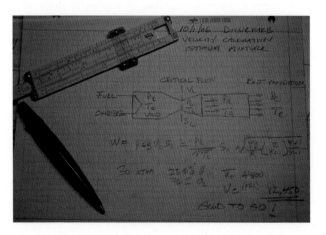

I think they understood that respect is something that flows in both directions and is really all most people are looking for in an equitable exchange. If you can earn the respect of the shop people, they will truly bleed red blood for you when the chips are down and you need it.

Engineers and metalworkers seem to come from different sets of molds. Understanding the basic differences goes a long way toward understanding one another. And as with most things, I happen to have an opinion. Engineers and designers are created for the most part in schools with their final luster coming from their first character-building jobs. Metalworkers, on the other hand, have learned a smaller part of what they know in schools, but the bulk of their career knowledge is learned on the job.

The world of the engineer is a much more open and collegiate environment when compared with the typical shop surroundings. They ask questions of one another in meetings and review each others work, looking for errors without placing blame. It's a more open and cooperative environment, much like the schools that produced them. It's okay to say, "I don't understand" or "I'm not following you." These are obviously fairly gross generalizations but you get the idea.

The world of the metalworker is much different. Many metalworkers have a modest-to-medium schooling background. When a metalworker makes a mistake or does not understand something out in the shop, the reaction is typically different. Somebody is usually upset and the individual who made the mistake takes the brunt of the blame squarely on the nose or, worse, their workmates relentlessly remind them of the specifics of their ignorance for the rest of their natural lives. The typical reaction is to hide or minimize all errors and mistakes. It's simply a matter of survival.

A metalworker's knowledge and special skills are the measure of self worth. It is quite normal for shop people to keep secret the specifics of their skills and tricks. This becomes the gauge of their value and standing in the shop pecking order. Shop people

quantify their performance in physical tons of completed work on the pallet and the bottom line on their paychecks. Kind of like the first tribe to figure out how to make fire. For a while, they were without peer and at the top of the heap. Then some missing link pre-engineer scratched a diagram in the dirt of how to make fire and all the fun was over.

This behavior can be seen throughout history. Engineers and designers write books, take notes, and make drawings. Metalworkers pass their knowledge on to the apprentice in the traditional manner by demonstration and lots of yelling.

So when you go out into the shop with this little insight into the kooky mind of the metalworker, use your powers of observation and see if you agree with me. Now off you go into the cold smelly shop and don't let the door hit you in the butt.

You will have to spend some time out in the shop to appreciate some of the problems that face metalworkers. Your standing in the eyes of the shop people is directly related to how much time you spend in their world.

When working on a new design, talk to the shop people who will be doing the work before you get too far along. They will see things you may never consider.

Promptly return any borrowed tool no matter how small or seemingly insignificant. They should all be boomerang brand, and always come back. I cannot overstress this point. After all, a 1/8 allen wrench is no reason to find yourself scratching the inside of a rough wooden box in a shallow grave under a lonely freeway overpass, now is it?

Sometimes it's much easier to tell the shop what you want than to try to over-specify or detail it exactly. Press fits, sliding fits, and threads are a good example.

Impress the shop with a great hand sketch (Figure 3-1). Nothing screams ability more quickly than an excellent sketch done on the fly.

Do your tolerance analysis and be realistic about your requirements. This area marks one of the great all-time abuses of shop resources. If you don't have time to tolerance with this kind of

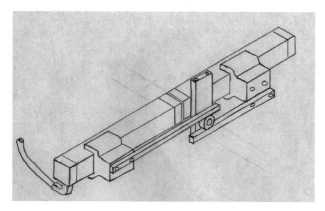

Figure 3-1: A clear hand sketch demonstrates your abilities.

thought, then leave them off all together or at least ask the shop what can be done realistically.

Allow cleanup cuts on stock sizes if the design will allow it. Your related tolerances should give the fabricator a choice. A plate that is nominally .50 thick might have a note, .485 minimum. This allows up to a .015 cleanup if the raw material is rough or has come in slightly undersize. If the drawing calls out carelessly .500 with a title block tolerance of ±.005, then your part just tripled in price.

Include stock material variances in your tolerance study. Half-inch thick plate or bar stock is rarely .500 like I see on almost every drawing that crosses my desk. Most of the time the engineer or designer has no intention of machining or surfacing stock material. A quick note next to the dimension .50 (STOCK) is a great way to communicate the proper intent.

Only ask for maximum speed and effort when you really need it. There is no faster way to wear out your welcome than to abuse this powerful tool.

Respect the shop's time. You might be in for a big surprise when your friendly yak time gets billed to your project.

Don't try to help out in the shop unless asked. Metalworkers are a bit territorial and not only bark, but occasionally bite.

Your drawings and instructions are your calling card out in the shop. Make them look good or, better yet, perfect (Figure 3-2). Lousy drawings and bad instructions are a hard reputation to break once you get it. I had one top notch designer who offered me lunch anytime I could find a mistake or omission on any of his work. It was five years before I collected.

Include the shop in critical manufacturing decisions. Having extremely specialized material and

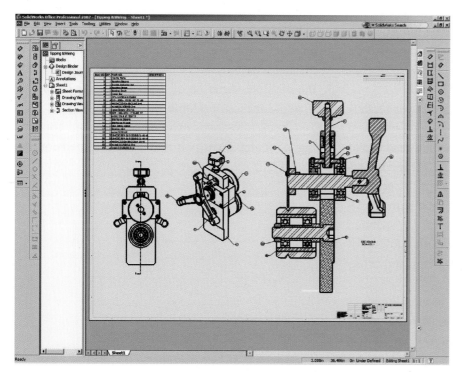

Figure 3-2: Top quality sketches help your work and your reputation.

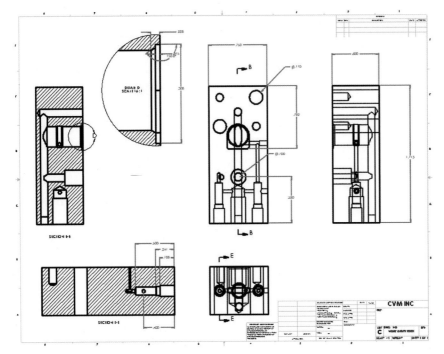

Figure 3-3: Using a breakout detail to clarify complex geometry.

process knowledge, they will spot things that are of concern from their viewpoints.

Have a regular presence in the shop. I know it can be cold and smelly, but it's easier to get answers to your questions when you take more than a passing interest. You may have had a semester of shop experience, but most shop people have decades under the bridge. Use this to your advantage.

Be sure to include the hidden lines in your different drawing views. Turning off the hidden lines is a trend that I have noticed since the widespread use of solid modeling. You may think it looks clearer but it's more like a game of "Try to guess what I'm drawing wearing a blindfold while standing on one foot."

When important information must be included in note or text form, attach a leader flag to one of the pertinent object lines in the drawing and reference the note in it.

Instead of dimensioning to hidden lines or features, try using a section view or breakout detail view (Figure 3-3). This is particularly easy if you are solid modeling.

Add an additional drawing sheet instead of cluttering one drawing page with a million overlapping details.

Include the dimension to the theoretical sharp intersection of any angular relationship (Figures 3-4 and 3-5). For that matter, dimensioning any acute angle or knife edge is a pretty dodgy thing, unless the angular tolerances are such that they include the possible variation of the actual knife edge. Try to avoid these if possible.

Check tolerances extra carefully when converting or using dual-inch and metric dimensioning.

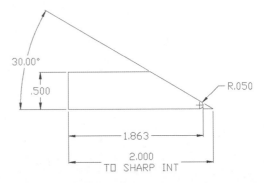

Figure 3-4: Include dimension to theoretical sharp.

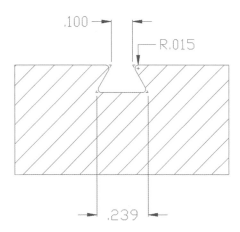

Figure 3-5: Include dimension to theoretical sharp.

There is a big difference between .005 inch and .005 mm.

Put the o-ring size numbers on the drawing near their respective grooves. Many times the shop has to test or leak-check a part. Having this information readily available on the drawing speeds things up. This information also communicates the grooves usage as a seal surface to the shop.

Use double dovetail o-ring grooves instead or single dovetails if you must use a retentive type groove. This gives the shop more flexibility and options for producing the groove. Single dovetails require fancy footwork on the mill or minor pain in the lathe. Some parts don't lend themselves well to lathes.

Go easy with welding. Massive over-welding is a very common and expensive practice (Figure 3-6). The complications from over-welding are time consuming to correct, let alone the extra time required to do the welding. How much is too much? One quick rule of thumb is a weld with an effective throat equal to the material thickness is equal in strength to the material. Anything bigger

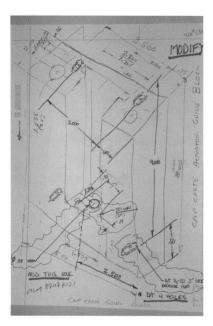

Figure 3-7: This design is hard to read.

than this is overkill. This is a generalization. The intended usage and design obviously play a part. This is why you went to school for all those years — so you could figure this stuff out.

If you make drawings like the fine example in Figure 3-7, don't complain when you don't get what you thought you were asking for. Looks like somebody spilled their spaghetti sauce on it.

Figure 3-8 shows a better way to round the ends of links, tabs, and eyes. A full radius is very sensitive to part width and looks lousy if it's not done perfectly. Two radii are simpler and easier to make look good.

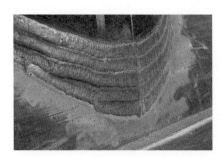

Figure 3-6: Big weld.

Figure 3-8: A different way to round tab ends.

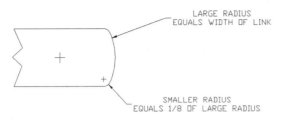

Figure 3-9: Measuring radii.

The large radius is equal to the width dimension of the link. The two corner radii are approximately 1/8 the size of the large radius (Figure 3-9). These dimensions are not anything magic, but are offered only as a way to get consistent results with all the different sizes of tabs that come up.

Try ordinate dimensioning for radial patterns (Figure 3-10). This can save setup time and shop calculation errors. X and Y coordinates are more accurate than angular layouts for large diameter patterns. Most of the time the shop converts angular specifications back to ordinate anyway. This conversion process introduces yet another chance for error.

For turned parts, the best dimensioning method is an ordinate scheme with all the Z axis features dimensioned to the far left side of the feature. Take a look at Figures 3-11 and 3-12 to see the difference. The image in Figure 3-11 is a jumble of dimensions with the machinist left to sort out by calculation what they need to make the part. The reason for using ordinate dimensions on lathe-

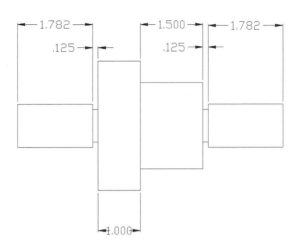

Figure 3-11: Dimensioning not optimized for lathe operator. Math required.

turned parts is in how machinists reference their tools. Typically they are touched off on a freshly-faced end which becomes the datum for the rest of the features. The type of dimensioning shown in Figure 3-12 can save time at the machine, with less calculating and more metal removal. For two-sided parts, you can have two datum ends. This style of dimensioning is all quite simple with electronic drafting, so why not give it a try.

Orient the drawing of turned parts in the same orientation they will be in the machine (Figure 3-13). On the same note, if possible, orient rectangular parts with the long axis running right to left. If you need to pick a datum corner, use the upper left hand

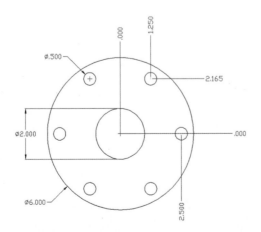

Figure 3-10: Ordinate dimensioning for radial patterns speeds setup.

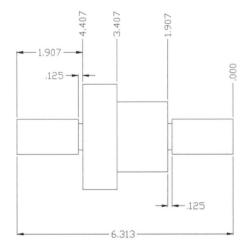

Figure 3-12: Dimensioning optimized for the lathe operator. No math required.

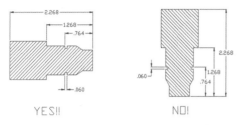

Figure 3-13: Drawing of turned part oriented for ease of manufacturing.

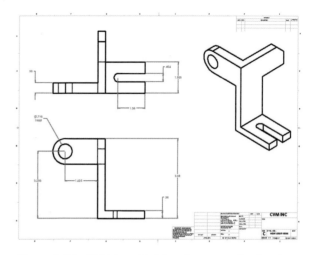

Figure 3-15: Right hand left hand as separate drawings.

corner if it makes no difference to your tolerances and features (Figure 3-14). The second-best choice would be the lower left corner. In general, we like to have the dimensional datum corners butted against our fixed reference surfaces on the machine.

Go easy when the inevitable mistakes are made. This is a double-edge sword that cuts deeply when misused. The next time you feel like blasting somebody over a mistake, think about having your drawings scrutinized for the next ten years by a pack of wolves on the scent trail of an engineer hauling a sack of bacon-flavored drawings in his briefcase.

In the modern electronic drawing age, it is a rather simple thing to output a complete drawing for a right-hand, left-hand, or mirror situation (Figures 3-15 and 3-16). Many mistakes are created in the shop because of confusion when working with the "wrong hand view." Every person who handles the drawing after the designer adds the note, "Right hand shown, left hand opposite" is in a position to make an error and waste time. All must interpret the details of the undrawn

configuration. The minor savings in time in the drafting department is eaten up by the first person who has to read the drawing.

Never, ever scale the electronically drawn part to "fit" the title block for printing purposes. Accurate dimensions should always be preserved electronically. If you must scale something, scale the good-for-nothing title block to fit the drawing. Many times these drawings are transmitted electronically and are never printed. I have seen this done more times than I care to remember. Nothing like getting your parts back from the laser or waterjet cutter and wondering why they are so small.

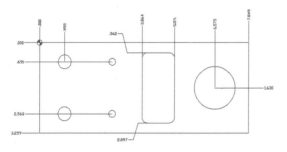

Figure 3-14: Part dimensioned using ordinate dimensions and upper left corner datum.

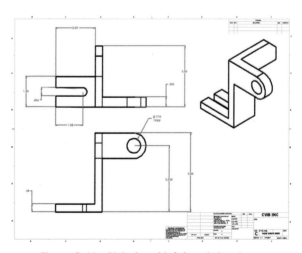

Figure 3-16: Right hand left hand drawings.

Be willing to listen and compromise if it makes sense. On the flip side, explain why you need something a particular way if it's important. A little information never hurt anybody.

Provide thread reliefs on OD and ID threads — typically at least one thread pitch as a bare minimum. You can't use a .015 wide thread relief with an 8TPI thread. The threading tool is 10 times this width just to make the thread depth correct.

Choose standard fine threads for custom-threaded parts if you have a choice or it doesn't really matter. Fine threads cut quicker than coarse threads. In tough materials, this can make a big difference. Along those same lines for custom threads, pick even numbers if it makes no real difference. On the lathe this makes threading a microscopic bit easier because of the thread chasing dial engagement points.

If you have the mating part for threading, machinists almost always prefer to have it available to double check their work.

I learned the hard way how difficult it is to sit in a drafting room designing parts or entire machines without the benefit of physical things in front of me for assistance. The next time you get mad at an engineer or designer take a deep breath and have a little empathy. These folks are just as hard working as any shop person. Office professionals are working with a different collection of information about the project than you are. What may seem like a simple change or solution may have gone through dozens of painstaking iterations in mind-numbing meetings and have much deeper roots than apparent from the shop. Cost, safety, and yes even political reasons shape the designs we see on the shop floor.

If you can find a good working balance between the shop floor and the office you are on your way to a much more rewarding time spent at your craft.

Shop Nickel Tour

The condition and maintenance of the metalworking shop have a direct effect on shop efficiency and morale. A bright, well lit, clean shop is easier and less mentally demanding than a wet, dripping, bat infested cave. If it looks and smells bad, it probably is.

An organized and well thought out shop space is a pleasure to work in. Having everything at hand and organized is like shopping in your favorite tool store. Retail hardware stores are a good example of what a nifty metalworking shop might look like. Good lighting combined with easy-to-find, well-stocked storage make the work run smoothly and efficiently. Don't underestimate the value of infrastructure improvements on efficiency and profits.

Floors

Light-colored floors are best for machine shops and areas where fine work is being done. They reflect light and give the shop a cleaner feel. It doesn't hurt that it's easier to find a lost part on a light colored floor.

Epoxy coating or even special-colored flooring tiles can be used to brighten the shop. Individual tiles are best because they can be replaced if they get damaged. The tiles also provide a little cushion if you drop a tool or delicate part. Plastic floor tiles in an array of colors are now available that snap together to make an excellent shop floor.

Welding areas need to be your basic plain concrete. Epoxy coatings or paint take a beating when hot metal lands on them. Depending on the class of welding work, paint or tiles may be acceptable.

Figure 4-1 shows a heavy-duty, steel working floor with cast iron platens set at floor level. These are used for fixturing and setup of large steel structures.

Smooth finish on the concrete makes sweeping easier and doesn't wear out the toes of your boots as quickly when you're crawling around on your hands and knees. Skip the non-slip sand and abrasive material in paint and epoxy coatings. At least sample a test section before you commit several thousand square feet into sandpaper. If you need proof, try sweeping the parking lot for a half hour

Figure 4-1: Steel working floor.

45

Figure 4-2: Wood brick floor.

Figure 4-3: Interlocking plastic floor tile.

and see what you think about roughened surfaces as shop floors.

One of the best floors that I have seen as far as durability and working comfort is wood. The little extra cushion provided by wood makes long hours on the feet and knees easier. It's not always practical, however, in the modern, tilt up, concrete shop buildings.

In many old time shops, you can still see wood brick floors (Figure 4-2). They are made up of thousands and thousands of 4 × 4 or 4 × 6 blocks cut off three or so inches long planted in the floor end grain up. Some of these floors are 50+ years old and still going. Cheap material and easy to fix if they get damaged. Figure 4-3 shows an interlocking plastic floor tile. These tiles can be snapped together quickly to create a great looking floor in no time. This type of tile has the advantage of simple installation. Furthermore, you have the ability to take them with you if you move.

Light

Every shop needs light. Standard warehouse lighting just does not cut it. The finer the work the more light is needed. If you're planning a shop and an architect tells you how many lumens per square foot you should have for a machine shop, be sure to upgrade to the next level. The architect may be

reading a book that was written shortly after the earth's crust cooled and people worked in caves using mammoth blubber lamps. Things have changed since then. Adding this extra light will save you in the long run from cobbling in extra fixtures because you don't have enough. Have you ever been in a shop with too much light? Your model here is the well-lit hardware store or other retail space. The best lighting is equal in intensity from any direction. Not always easy to do, but a good goal. A bright, well-lit shop is a simple morale building tool.

Modern efficient high intensity fluorescent is best, followed by Mercury vapor lamps; they may need to be mounted high to clear crane rails. Stay away from sodium vapor because is casts a sickly, yellowish light. Mercury vapor lamps take a while to warm up, so if you need instant light, fluorescent will be your best bet. In shooting some of the pictures for this chapter, I had trouble with the lighting in one shop that had sodium lights, which required some color correction.

The lights should be mounted fairly high to miss the tops of machines. I have seen many bulbs broken in low-hanging fixtures from flying parts and from handling long materials in the work areas.

Task lighting should be easily positioned and, ideally, cool running. Skip the food warming flood lamps. It takes it out of you to have a light

Figure 4-3b: Make food easy to find.

that could cook a hamburger beating down on the side of your head all day. Magnetic bases can easily be moved to new positions. Several excellent cool running LED machine task lights are now available.

Food Preparation Area

Every engine needs fuel and every shop needs a refrigerator and at least one microwave oven. If you have a lot of workers, consider having two or three microwaves. This eliminates long lines and folks popping their stuff in before the lunch bell to beat the logjam for a single microwave. There is always one person in the shop who has to heat his fermented mackerel and sardine casserole to the temperature of the surface of the sun and permanently contaminate the only available oven. If you have two ovens you can designate one for questionable food items and forensic leftovers.

The refrigerator should be cleared out once a month or, better yet, once a week. Stuff tends to accumulate there and be forgotten. Get rid of anything that looks like a failed science project or hides when you open the door.

Food Consumption Area

One of my personal favorites. The lunch room or break area should be separate from the actual shop. It's nice to sit down and not hear the CNC machines still running while you're taking a break.

A large table promotes camaraderie and boosts morale. Get a couple of marginally work-related magazine subscriptions and leave the magazines in the lunch room.

Heating and Cooling

Many metalworking shops start out life as warehouse spaces. They are typically cold in the winter and hot in the summer. If you have a choice, insulate and then add dedicated heating and cooling. Typically these areas need to be kept at different temperatures than office spaces. Be sure they have their own climate controls. I don't think you would get much complaint from anyone if the temp was kept at a year-round 68 degree F. This also keeps your precision measuring equipment closer to their calibration standards.

Workbenches and Tables

Several types of specific workbenches and tables are needed in any shop. All should be of sturdy construction. A workbench that wiggles or is rickety is very annoying. The basic types are basic work and assembly, welding, and mobile. All three have special requirements. Many smaller shops on a budget built their own when they have slack time.

Basic Assembly and Work Table

This table should be rectangular in shape with a maximum width of 36 inches. Much wider than that, and its hard to reach across without walking all the way around.

We surface our general benches with a white board material similar to what is used for dry erase boards (Figure 4-4). It's sometimes called melamine at our local home center. We add a short rim on three sides on some of the benches near the machinery to keep tools and parts from rolling off the back or sides.

The table's height should be 36–40 inches for most folks. Lower gives you back pain after a long

Figure 4-4: A rim will keep tools and parts from rolling off the bench.

Figure 4-5: Mobile cart.

day. Generally, higher is for fine work and lower is for heavy work. Light-colored replaceable tops reflect light and let you see small parts easily. When they get beat up, change them out and they look brand new. Beware of the lower shelf; it's an area that tends to collect junk.

I used to believe that the only thing that should be on wheels is the hand truck and the forklift. I have since reversed my thinking completely. For maximum versatility most, if not all workbenches should be on wheels, with the exception of heavy welding tables or benches that have machinery attached to them. These need to be very stable to do decent work. Let me clarify this a bit. The wheels need to be swivel with a brake and a swivel lock. If you don't use this type of wheel, skip putting wheels on a workbench. The swivel lock, sometimes called a total lock, is the key ingredient that keeps the workbench from moving around until you want it to.

Having most of your workbenches on wheels allows you to re-configure your workspace quickly for any job that comes in. Such workbenches are easier to move for cleaning purposes and can be moved if more floor space is needed for a large project. For jobbing shops that never quite know what will crawl through the door, or what the customer trots in, bringing the table to the job is valuable.

Mobile Workbenches and Carts

These are much smaller than a regular workbench (Figure 4-5). These are used to move raw material and parts between machines and processes. Good wheels are a must for these often overloaded carts. Small is beautiful with this type of cart because of where they are used. They are easy to stow and push around because they hold only a couple of hundred pounds at most. We also use furniture dollies for moving heavy plates and boxes around to the different work centers (Figure 4-6). They are cheap and roll well, even with a heavy load. The carpet on the surface keeps your stuff from sliding off and they store relatively flat.

Figure 4-6: Furniture dolly.

Air Supply in the Metalworking Shop

The shop air supply is a critical path in the modern metalworking shop. If your air system goes down, it can affect your entire operation. Many CNC machines require air at a certain pressure and volume to run. A high-quality, properly-maintained system is crucial.

Shop size and machinery dictate the size of the compressor. Be sure to engineer in room and branching possibilities for future expansion when choosing your piping system, compressor, and storage tank. It's easy to put a tee in instead of an ell to allow future expansion opportunities.

Not the sign you want to see on the first day of a new welding job (Figure 4-7)! I hope they drained the water from the air system this week. . . .

Consider multiple smaller strategically-located air tanks instead of one single large storage tank. Often it's easier to find precious floor space for smaller tanks.

The distribution network should be steel or copper pipe. PVC works and is safe under most settings, but has the cheapskate rookie look to it. OSHA says anything below eight feet off the floor needs to be metal, so you might as well make the whole thing out of metal. Copper gets my nod of approval for corrosion resistance and low leak potential.

Figure 4-8: Simple drain cock.

When in doubt, place a valve. Each drop and dedicated air line to a machine should have a valve. Depending on your shop layout, you should have the ability to isolate areas for maintenance, repairs, or expansions without having to shut down the entire air distribution system.

Simple drain cocks should be in the bottom of every drop in the shop to bleed accumulated water (Figure 4-8).

Dedicated air nozzles on machines should be plumbed to the hoses without quick disconnects (Figure 4-9). This prevents the unwanted removal and inevitable wandering of the air nozzle and the time wasted hunting one down.

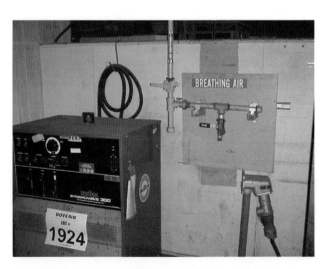

Figure 4-7: Air supply is essential to the shop.

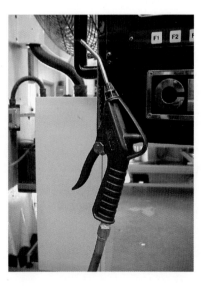

Figure 4-9: Dedicated air nozzle.

For whatever reason some drops seem to collect more water than others. It's great if you have a chiller dryer on the compressor. But very few air systems I have used were completely free of water. Therefore, provide a way to get the water out and use them diligently.

Assign the duty of draining the compressor to one of the apprentices on a regular basis, typically at least three times a month. If the compressor is used heavily, invest in an automatic drain.

The air supply to the air nozzle should be regulated to a lower pressure than the main shop air. It cuts down on the raw noise and is safer against bare skin. Over time, this noise on unprotected ears will cause hearing damage.

The hose for a dedicated nozzle should be optimized for length. A long coil of unused hose lying on the floor collects chips, coolant residue, and junk; it's hard to clean around, let alone the leg snaring abilities of simple rubber hose. For air blower nozzles, a small inside hose diameter (1/4) makes for a lighter, more flexible hose. I have never had a problem with flow for an air nozzle using this size.

Air Tool Supply Lines

These lines are sized for each tool. Use your largest air hog to determine inside hose diameter. If you're doing delicate work will small tools you can make up a short whip of hose that has a smaller inside diameter that connects to the main hose. Make these a maximum of six feet long. This keeps the large main supply hose from dragging your arm down all day long. On a couple of my tools, I have plumbed the whip right into the tool body to eliminate the bulky connector or swivel joint.

Hose reels are great for neat clean storage of air lines (Figure 4-10). Put two at opposite ends of the shop and have them slightly overlap in length with the hose fully extended. Don't forget to put a universal female coupler on the end. Make sure that it sits higher than the tallest guy in the shop's forehead, unless you like the knurled forehead look.

Figure 4-10: Hose reels help store air lines.

Don't let the hose slide through your bare fingers when retracting the hose reel. Razor sharp chips get stuck in the surface of the hose and will slice your hand. Instead, keep your hand on the quick disconnect at the end. Pain awaits the daydreamer or unwary with a bad bashing.

Be sure to include a valve right before the hose reel so you can service the reel.

Don't drive forklifts and other heavy-wheeled machinery over light-duty air lines. Otherwise, you will shorten their life considerably by driving chips and scrap metal into the surface. I got yelled at for this many times as a teenager in the shop, so now I get to pass it on.

Air tool oil should be readily available near air supply points. This accessible location encourages the lubrication of expensive pneumatic tools. Actually it encourages a convenient storage place to keep the **empty** bottle of air tool oil, but that's another story.

Provide simple professional looking hooks for storing air hose extensions (Figure 4-11). A similar hook works well for getting extension cords off the floor. Nails and spikes are for carpenters and other woodchucks, your hoses and cords will thank you with an extended life.

Figure 4-11: Hooks help store air hose extensions.

Figure 4-12: Universal female quick disconnects.

Use universal female quick disconnects (Figure 4-12). These fit several of the most common types of male plugs encountered in air systems. They save time wasted hunting down the correct fittings or adapters. It seems like every shop independently decided to use a fitting diametrically opposed to every other shop's fitting. I have a little box in my toolbox with every adapter under the sun to prove this.

Raw Material Storage and Handling

Every shop must store some raw materials. The trick is to store them in such a way that you can get to the piece you want with the minimum of human effort. Labor is typically the most expensive

component of any job. Anything you can do to shorten the labor path during a job will have a short payback period.

Cut 20-foot bars in half for easier handling. It depends on the type of work a shop does, but rarely are full 20-foot bars utilized without cutting. In small shops, 20-foot lengths of heavy materials can be extremely difficult to handle and tie up expensive labor to cut off a short length. This challenge goes on and on until the material gets short enough to be handled efficiently. It holds true especially for machine shops using short fat chubs. The time to cut the bar is when it comes in the door from the supplier. I don't know about you, but I'm too old to buck 12-foot lengths of 3-inch round out of the bottom of the rack.

Figure 4-13 shows a cart we made for moving bar stock around the shop. It is the correct height for the horizontal band saw so material can be pushed right into the saw as soon as it's unloaded. The support tubes are also set up so you can load and unload it with a forklift right off the delivery truck.

Graduate your material drops by length, for example, full bars, partial bars and stubs. Most people are lazy and will try to use the material with the easiest access before pulling out the full length materials. If you provide and maintain a system for finding the optimum stock in the speediest manner, you shorten the job. This is proven time and again when we filter the racks

Figure 4-13: A cart helps when moving bar stock.

Figure 4-14: Racks and bins provide additional storage.

Figure 4-15: Identify your materials.

and bins (Figure 4-14). The size of the keepers just keeps getting smaller and smaller until they just disappear.

Weed and organize the material storage once in a while. This is a good job for apprentices and helpers. They will get acquainted with the shop's raw material inventory. This task is a great way to give them direct contact with many different material types, sizes, and forms they will need to know about later as their careers progress.

Beware of packrats. Machinists and welders are natural scrounges and packrats. Want proof? Look under the workbenches and in all the nooks and crannies in their work areas. Some materials and leftovers really need to be scrapped for running an efficient shop.

Get rid of materials that cannot be easily identified. If you don't know what it is, how can you

use it? If you must be a packrat and save it, mark it unknown. "When in doubt toss it out."

Certain form factors of scrap and material drops are virtually useless. Long thin strips of sheetmetal and triangles that are cut off from almost anything are two examples that can hit the scrap bin right out of the gate.

Almost all circles and discs should be saved. Unless your processes generates a large quantity of disc-shaped parts, these should be saved.

Material identification is a challenge to maintain in a job shop. Each time a piece of material is cut, there is a chance to lose the identity trail (Figure 4-15). It's every person's responsibility to make sure the identity chain does not get broken.

Keep marking tools readily available in the areas where materials are stored and rough cut.

Put the identifying marks on the end surfaces of bars. Be sure to mark both ends.

Engrave or mechanically stamp the material type directly on the bar end (Figure 4-16). Tags fall off and ink smears. Tape or labels are next to useless. They are placed on the side of the material where other material sliding in the rack can obliterate the identification. The sticky labeling goo also has to be removed at some point by somebody on the payroll. I hate label goo on my consumer packaging, so why would I want it on my raw materials. When you mark the end surfaces of material, it fits in collets and vises; the

Figure 4-16: Indicate the material type.

Figure 4-18: Use tubes to store small rounds.

Figure 4-19: Storing other small materials.

first cut in the machine takes care of removing the engraving.

Don't use obscure internal company terminology or secret codes for common materials. Use the same terminology the suppliers use. For certified materials, include the purchase order number the material was bought with. This information helps when tracking down copies of test reports and conformance information.

Store flats on edge in the vertical orientation (Figure 4-17). Graduate them by width to make inventory and removal easier. If you don't store bars this way, the one you want is always on the bottom.

Store your most accessed materials at waist height. I always hate it when I get my eye core sampled by a piece of tubing when I'm reaching over my head for something that should be lower.

Metal supplier color codes are unreliable. There is no real color standardization in the metal supply

industry. These should only be used as a generalization or identification when combined with another method.

Small rounds should be stored in tubes so they don't slip between the rack dividers (Figure 4-18). Alternately you can form simple sheetmetal trays for the slots that contain small materials (Figure 4-19). This keeps the more flexible materials from drooping and missing the rack supports.

Material Identification and Characteristics

Machinists and metalworkers need materials from which to make things. There are literally hundreds of different materials that are considered common nowadays. When you think about all the different kinds of operations that a piece of

Figure 4-17: A good method for storing flats.

material might be subjected to during its fabrication, the knowledge required to keep track of the different characteristics is daunting. There is not one go to place to learn these things; unfortunately, most are learned the hard way.

At one place I worked, I was subjected to six distinctly different materials during my first two days. Each had a specific set of do's and don'ts. For the modern metalworker, a solid knowledge of materials and their qualities goes with the trade. When you are given an unfamiliar material to work with, take a quick minute to look up or at least ask about the common characteristics and problems. Some materials are so sensitive to common shop chemicals and substances that they can be permanently damaged without you doing anything intentionally. "When in doubt, check it out."

The ability to identify different materials at a glance by their look, feel, and mechanical qualities takes time to develop, but the practice necessary to learn is well worth the effort. Any given shop will have a varied cross-section of different materials—several dozen at least—that the metalworker meets. How many times have you picked up a piece of unmarked material and wondered what exactly it was? Here are some quick, shop expedient methods that can be used to narrow the field of doubt.

The human hand and eyes are extremely sensitive comparative instruments, if you learn to trust them. Properly trained, they can discern minute differences in color, thermal conductivity, ductility, density, and modulus. These are not absolute methods, but they will help you narrow the field and tip the scales in your favor when examining and choosing different materials. The more you know about the different materials you work with, the more you will be attuned to each of their special qualities. This intimate knowledge of materials is especially useful when trying to build or design parts with different functional requirements.

Always try to make your comparisons with a piece of known material in hand. It will improve the accuracy of your observations. If you think

something is 7075 aluminum, have a piece of known 7075 while checking the unknown material.

Don't rely on any one characteristic. Use as many as you can compare to remove doubt.

Color. Aluminum and steel have distinctly different colors. Both are silver, but aluminum has a slight bluish tinge that steel lacks. The color difference between steel and stainless is less obvious, with stainless having a slightly more silver color and steel more toward grey. Copper is redder than brasses or bronzes and brass is more yellow than most bronzes.

Magnetism. The small magnet on a pocket screwdriver (Figures 4-20 and 4-21) can tell you a lot when you're hip deep in the metal rack or you see a funny looking dowel pin in the box of stainless pins. I always have one of these cheap magnetic screwdrivers in the pocket of my apron. It's my combination pry bar and metal identifier.

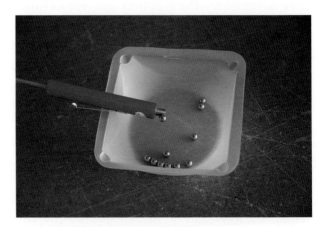

Figure 4-20: Using a small magnet.

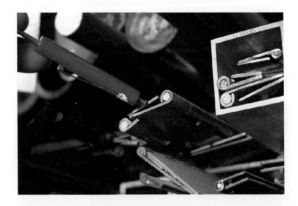

Figure 4-21: Pocket screwdrivers have many uses.

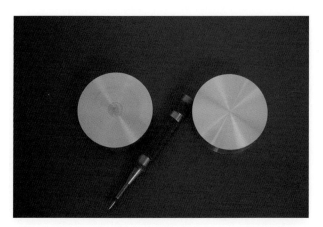

Figure 4-22: Distinguishing between materials of different hardness.

Density. Many materials can be identified by their density or lack of it. 7075 aluminum is noticeably heavier than 6061 for an equal-sized part, even at only 3% heavier. This density, combined with higher hardness, makes 7075 stand out from the more common 6061.

Hardness. Different types of aluminums can be detected by the differences in hardness. This also works well for many plastics. For example, Delrin is harder than ABS or polyethylene. Acrylic is harder than polycarbonate. A simple test with an automatic center punch (Figure 4-22) can distinguish between two materials of different hardness. In the example shown in Figure 4-23, the sample on the left is 7075 Aluminum and the sample on the right is 6061. The center punch mark is slightly larger in the 6061, which is what we would expect to see in this case, knowing the characteristics of the two materials.

Chip forming. Each material has a characteristic chip formation. For example, 303 stainless tends to short chip when compared to 304 or 316. Delrin makes short chips which smell different than ABS or polyethylene during cutting. Most plastics have distinctly different smells when a small sliver of the material is heated. Caution: some fumes from heated plastics are detrimental to your health.

Thermal conductivity. Aluminum feels warmer to the touch than stainless of similar finish.

Mechanical condition. Harder material with higher tensile strength will ring at a higher pitch when dropped or tapped with a tool.

Get your hands on an armload of metal and plastic supplier catalogs (Figures 4-24 and 4-25). These provide a wealth of useful information on material properties and processing. It's easy to flip back and

Figure 4-24: Metal and plastic supplier catalogs.

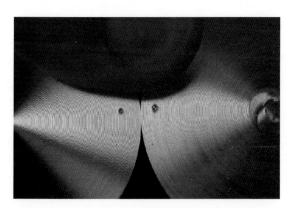

Figure 4-23: Comparing 7075 and 6061 Aluminum.

Figure 4-25: Supplier catalogs provide valuable information.

forth in these great little books comparing the different qualities. Once you have some experience in the shop with some of the materials, you can easily correlate the observed differences with the book data. I also keep a big binder that I call my "Big Book of Knowledge." It has all the data sheets and information I have had to look up in the course of doing my job. Whenever I get a nice piece of information on, say, the factory flatness standards of stainless sheet metal or the corrosion qualities of Titanium in nacho cheese sauce, I put it in the book. It has been extremely handy over the years because it relates directly to personal experiences.

Elongation

Elongation numbers gives us a clue about the formability and ductility of a material. The higher the percentage, the farther it can be worked.

Modulus of elasticity. This measurement tells us about a material's stiffness. The higher the number, the stiffer the material. Stiffness and strength are two very different qualities. Don't get the two mixed up.

Yield strength. This measure shows the value we use to determine the upper end of basic strength differences between materials. Ultimate strength is not as useful as a design parameter because most materials have already permanently deformed or yielded long before it sees the ultimate strength. These numbers just give us a relative comparison of strengths. Yield strength shows us at what point the material will be dimensionally altered.

Toughness. This indicator measures a material's ability to absorb energy. Impact and notch test numbers give us a way to compare different materials. Higher numbers mean tougher material. They also mean more sweat if you have to apply human energy.

Thickness

Sheet is anything under .188 thick. Anything thicker is called plate.

Sheets and plates have bow (parallel to the rolling) and camber (perpendicular to rolling) tol-

Figure 4-26: Cast aluminum tooling plate.

erances that would surprise you. Most of the time, these materials come in tolerances better than what is allowed, but sometime they don't. These specifications are available from your suppliers. If they don't want to give you the information, take your business elsewhere. Use this valuable information for your real world design efforts and tolerance analysis.

Cast aluminum tooling plate sold under various brand names is not ground flat. It is cast against a ground surface, which gives it that ground look that sells more plate (Figure 4-26). This material is not particularly flat by machinist flatness standards. First get the specification sheet from your aluminum suppliers. Then get ready for a surprise when you see how much flatness deviation they graciously allow themselves. Keep in mind that many of the cast varieties of tooling plate offered are not repairable with welding. There are only one or two that I know of the welding is even recommended. Screw ups can be difficult or impossible to repair. The major redeeming factor is that large amounts of metal removal have very little effect on geometry.

Aluminum sheet is specified by decimal thickness, not gauge number. The salespeople will generally politely overlook this common oversight because they want to sell you material and not make you look ignorant.

Order your sheet materials with PVC covering to protect the finishes (Figure 4-27). The small

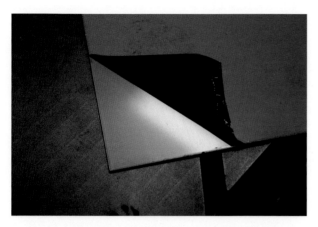

Figure 4-27: PVC covering helps to protect finishes.

incremental cost is repaid in less finishing work. For finishes produced in the shop, we sometimes protect the surface with clear shelf paper (Figure 4-28). It is readily available at any hardware store and will protect the finishes you create as they move through the shop. We recently started using a similar protective sheet material called Surface Armour. It is available in many different levels of tack, depending on the material you want to protect and how long you need to protect it. It is available in rolls in any width you need. See the suppliers' index for contact information.

Carbon steel and stainless steel sheet of the same gauge number have different thicknesses. To avoid confusion, its easier to call out the sheet-metal thickness as a decimal dimension with a tolerance on a drawing. Be sure to include an

Figure 4-28: Clear shelf paper also protects finishes.

appropriate tolerance to the thickness to encompass the manufactures gauge tolerance.

Beware: plastics typically have a thickness and diameter tolerances of $+/-$ 10%. This has caught many people by surprise because of its large possible variance.

All sheet materials have a thickness tolerance from the manufacturer. Be aware of this tolerance to avoid mistakes and to use to your advantage. This tolerance shows itself when doing precision forming where the thickness has an effect on the forming operation.

Sheet materials are usually thicker in the center of the sheet parallel to the rolling direction than near the edges.

Sheets are always slightly oversize in length and width. You can never count on the accuracy of the sheet width or length. There is a reason they sell materials by the pound—all the extra weight goes on your bill. If you really need an exact width, plan on re-squaring the material.

Galvanized sheet is hot rolled sheet that has been pickled in acid to remove the mill scale before plating. It is softer than cold rolled sheet of equivalent thickness. It's handy to keep around for durable templates and things you don't want to paint.

Material Lengths

Bar materials come in many different lengths, depending on material type and form. When calculating material needs, be sure you understand how the material will arrive so the proper allowances for yields can be figured. I do not claim to understand why the suppliers seem to hold on to these oddball differences. It sure seems like the industry could standardize and simplify this.

Steel pipe and tubing come in 21-foot random lengths. Okay I get this one, twenty feet plus a little.

Extruded aluminum bar shapes come in 12-foot lengths. Not sure about this one. Maybe is has to do with droop when it exits the extrusion die.

Extruded aluminum tubing comes in 20-foot lengths. Whoops! There goes my theory on droop.

Aluminum angle comes in 24-foot lengths. This is to beat the pipe folks.

Cold rolled steel bars come in 12-foot lengths. Precision cold rolling might explain the need for shorter bars. But many have sheared ends so they would appear to be rolled as longer lengths.

Hot rolled steel bars come in 20-foot lengths. Why not 21 feet?

Stainless steel bars come in 12-foot lengths, depending on whether the bars are true flat bars or Gauer bars, which are slit from coil. This makes no sense either way.

Now if you are calculating yields for a single length of part, this is all pretty straight forward. Part length plus a little for cutting, then divide this into the bar length. When you have an assembly like a large machine frame with many different cut lengths is when it can get a little tricky. With expensive or hard-to-procure materials, a few minutes spent looking at a decent cut list can save thousands of dollars. This is best handled by the person who would normally do the cutting.

Pipe Sizes

All true pipe of a given size will have the same outside dimension. The inside diameter changes with the pipe thickness schedule. This is a carry over from the good old days when most pipe connections were threaded. You only need one size threading die for each pipe size. Now, all it does it confuse purchasing agents who think everything round with a hole in it is pipe.

A couple of pipe sizes that make great telescoping assemblies are 1-inch schedule 40 and 1-1/4 schedule 40. This makes a great adjustable stand (Figure 4-29). For heavier duty stands, 2-inch schedule 80 and 1-1/2 schedule 80 make good telescopes also.

Square tube telescopes are more difficult to get right. Internal weld seams and lack of size choices limit this. An alternative is to add material to the outside of the inner tube to make a telescope.

Figure 4-29: Pipes can be assembled to make an adjustable stand.

Plastic strips can be attached to provide a bearing surface for the telescope.

Special Materials

Several special materials are worth mentioning here; you should get acquainted with them if you don't already know them.

Aluminum bronzes. These super-strong bronze alloys are a near perfect match for sliding contact with most stainless steels. Drilling, tapping, and reaming can sometimes be a minor pain. Tensile strengths can go higher than steels with some alloys.

17-4 Ph Stainless steel. This versatile precipitation hardening stainless has similar corrosion resistance to type 304, but can be hardened with a simple low temperature heat treat to RC44. It welds with normal processes without fuss. It is very stable in heat treat and can be machined easily after heat treatment. Unlike some of the other precipitation hardening stainless steels, this alloy is readily available in many forms.

Steel by the trade name Stress-proof otherwise known as 1144. This is an easy-to-machine, tough, strong steel. Over 100KSI yield strength as delivered, it can be further heat treated. It is a favorite of the screw machine industry. Its only real problem is that it not recommended for welding.

8620. Another strong, easy-to-machine steel. This alloy can be purchased most commonly in

Figure 4-30: 8620 is an easy-to-machine steel.

Figure 4-31: Producing a hard case with a tough core.

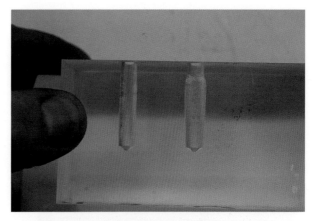

Figure 4-32: Spindle speeds can affect the quality of drilled holes.

rounds for turning work. It is easily welded by all common welding processes. The carburizing heat treat can be easily controlled to produce a deep hard case with a tough core, as seen in Figures 4-30 and 4-31.

Plastic Tips

When working with plastics, beware of the characteristics of these materials that make some operations difficult. Most plastics expand much more per unit length per degree than metals, up to 10× greater. The rule with plastics is to keep the heat out of the part. Plastics soften and melt easily under the friction of cutting tools. Try to use sharp tools and coolant when possible to reduce melting, and remove the heat generated by cutting. Plastics conduct and shed heat much more slowly than metals and can be damaged by localized heating. Slower spindle speeds and higher chip loads can help if you are having a problem. The difference

between the two demonstration drilled holes in Figure 4-32 is that the better looking one was peck drilled at a much slower spindle speed. Both holes used the same feedrate. The cycle time was marginally shorter for the bad hole, but who cares? It's a lousy hole.

For intricate parts with tight dimensional requirements, you may want to try stress relieving your plastics. Most heat treat shops can handle these requests. Be sure to ask your plastic supplier for the proper temperature and time profiles for your specific plastic. This can be done before machining and fabricating, or as an intermediate step prior to final machining. You can do this work yourself in an oven with a good temperature controller and timer (Figures 4-33 and 4-34). Typically, you heat to just below the heat softening temperature and hold for a time related to the part volume. Then reduce in controlled steps until reaching ambient.

Safety Equipment

Living is a dangerous occupation—just look at how many dead people there are. Metalworking is no more dangerous than any other human occupation; its hazards are just different. I believe that safety is a situational awareness issue as opposed to an equipment one. Just because you have protective equipment and guards in place does not assure any particular level of safety. On the same note,

Figure 4-33: You can stress relieve plastics in an oven.

Figure 4-34: The oven should have a good temperature controller and timer.

experience and training can help, but do not guarantee safety either.

If you never get in a car, its less likely you will be in a car accident. If you never stick your arm in a log chipper, you will most likely never have it removed by one. There are two kinds of hazards. The first are the ones you can just plain avoid. And the others are hazards that can seek you out.

I was at a garage sale one time many years ago. It was one of those once in a lifetime, super duper sales where a retired metalworker was selling off his junk pile to move across the country. A friend and I must have spent the better part of the day rummaging around in the various rooms and sheds in a machinery-drunken stupor. We each had a pile of pirate treasure that grew steadily as the day wore on. About the time when we were thinking about setting up our tent and camping out, we found a rich vein of ore in the garage and started rooting like a couple of truffle hunting swine.

Sometime in the late afternoon, a neighbor stopped by and asked the owner of this fantastic collection to help him with some metal-related project. Only later did I see that it was a mower blade that needed some kind of repair. I suppose this owner was the Mr. Fixit of the area and this neighbor, read the writing on the wall, figured he had better get it fixed now or never.

The owner had pulled out all the stops to help unload his massive collection of rust colored treasure. He had carted a large portion out of the basement and set it up along the driveway so more people could see the stuff at one time. I suspect it was so that people could actually get into the basement and move around. This guy had that much stuff. I get misty eyed just thinking about it.

My friend and I were near the actual garage at the far end of the driveway. The owner had set up this mower blade in a vise attached to a bench that he dragged out of the basement. Out of the corner of my awareness, I heard him hammering on the errant blade clamped in the vise. I paid it no real heed because I was at least fifty feet away.

After a vicious round of hammering, I found a real gem in the pile I was sorting through. At that moment I stood up and raised the bauble so my buddy could see my find and the gloat on my face. Bing. One more hit with the hammer. A sudden pain at the corner of my unshielded eye. Apparently Mr. Fixit was using a cold chisel to chip the heads off several steel rivets holding the mower blade to the hub. A chunk of rivet clipped me in the corner of my right eye. I was extremely lucky—the damage was only a small nick and a watering eye for a half hour.

This is an example of a hazard that seeks you out deliberately.

So what can you do about these two types of hazards? Avoidance is pretty effective, but doesn't get much work done. Because we cannot avoid all the hazards we are exposed to in our trade, the best form of protection we can use is awareness. Just like it's pretty hard to sneak up behind the karate grand master, we must have this same situational awareness to the hazards of our trade. We are the prey and the hazards are the predators. If something wants to eat you for lunch, then you should be paying close attention to what that thing is doing at all times. I have taken quite a few folks to the clinic to get stitched up because they let the predator have a little bite.

The hazards that seek you out are tougher to guard against. Hardhats, glasses, earplugs, and seatbelts are examples of things we can use to mitigate the seeking variety of hazards in our chosen trade.

Here are a couple of things that will save you a lot of grief later in life if you use them diligently. What are the most common traits you have seen with career metalworkers? Two that stand out to me: the majority of the old timers I have met are half deaf and three quarters blind, and all have bad knees. I can't help you with the knee problem, but the ears and eyes can be saved.

Ear plugs. The constant background noise in a metal shop destroys the hearing of most people after years of exposure. Unregulated blow guns and the whine of a geared engine lathe will detrimentally affect your hearing without a doubt. Earplugs are a hassle to get used to, but once you get past that break in period, you won't be able to work without them (Figure 4-35). The "roll up, throw away" plugs seem to be the easiest to wear and have a decent noise reduction rating. Skip the stupid earplug leash. If you take these out of your ears and put them back in more than once they are

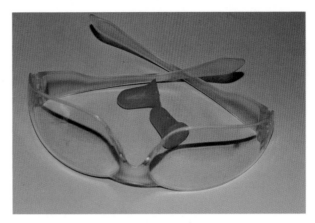

Figure 4-35: Safety glasses.

grubby from your hands or your wax-filled ear canals. All the string does is get snagged on every protrusion and handle in the shop, yanking the plugs right out when you least appreciate it. If I have to remove earplugs frequently for some reason, I switch to ear muffs.

Safety glasses. In any shop that does a significant amount of welding, glasses of some sort are a must (Figure 4-35). The stray UV light eats away at the peripheral vision of an unprotected set of eyes until your vision is ruined. Clear plastic safety glasses are effective against UV light. 99% of the harmful rays are reflected off clear plastic. Be sure your safety glasses have side shields. This is the area that our peripheral vision is attacked by the harmful UV light from arc welding.

"Safety glasses are for sissies, it takes a real man to face blindness!"

Scotch Super 33 Electrical Tape. This is the best industrial field dressing for cuts and scrapes. Look at this nice snug wrap (Figure 4-36). It looks more like it was painted on instead of wrapped.

Under normal combat conditions the prepared shop veteran expects a certain number of cuts and bleeds. A quick wrap of electrical tape keeps the bleed off the work until you can clean and dress the boo boo. It conforms to the curves of your fingers and stretches as you move. Also it doesn't leave the sticky residue that bandages leave, which you have to remove with acetone or some other nasty solvent. Try keeping a normal bandage on with your

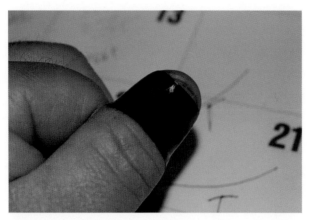

Figure 4-36: Scotch Super 33 electrical tape provides a snug wrap.

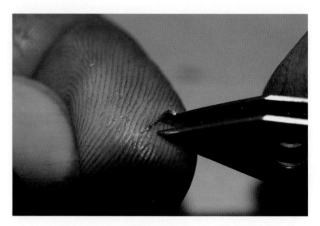

Figure 4-38: What I did for my readers!

Figure 4-37: Uncle Bill's sliver tweezers.

hands in coolant all day long. For these reasons, I always keep a roll of electrical tape handy in my toolbox.

Uncle Bill's Sliver Tweezers. Whoever designed these tweezers, got it right (Figure 4-37). They taper not to a deadly sharp point, but a small flattened point that meets with micrometer like precision. Those micro-fine annoying metal splinters don't have a chance with this fine tool. And yes, I really did deliberately stick a metal splinter in my finger for your reading enjoyment (Figure 4-38).

Tool Crib

My old teacher Doug told me when I asked him what tools I should bring into work, "Bring everything. Then as you see how the work goes, you can take the stuff you're not using home." He worked

out of a small sheetmetal toolbox that he had bent up at some point. The tools he had in the box were a collection of new and old. Several years before somebody had broken into the shop where we worked and stolen some tools. One of the boxes they took was Doug's. The part that was slightly funny was that these guys came in through the roof and then broke a window to get out. When the cops were there looking around and making their reports, Doug got one of his tools back. Apparently one of the tool thieves had used his 12-inch crescent wrench to knock the windows out. The cop handed it to Doug who then asked, "Did you check this for fingerprints?," to which the cop replied, "Oh. I guess not."

Doug extended the generous offer to me that I could borrow any of his tools so long as I was still buying tools myself. This is a basic ground rule and a great way to foster tool appreciation. If I stopped buying tools, there would no need for me to borrow any of his. You can judge a person's commitment to their trade by the investment they make in their tools.

If you borrow a tool, be sure to return it promptly. It is a major infraction in the metalworking shop to return borrowed tools lazily. It's very disrespectful. The white collar equivalent would be borrowing an engineer's toothbrush or pocket protector and failing to return it. I have found that engineers and scientist tend to view hand tools as objects instead of treasured personal possessions. Don't make the mistake and create an enemy by not returning a borrowed tool enthusiastically. I've seen some people pretty angry over an 1/8 allen wrench.

I advocate buying tools of the highest quality you can afford. If you're planning to make your living with them, the initial painful cost is amortized over your lifetime. Besides you are directly supporting somebody else who thinks enough of fine tools to actually make them. Any dedicated craftsman appreciates and uses fine tools as part of his or her trade.

Part of the fun of doing the work is using the tools. It's not much fun if you pick up a tool that you use everyday and regret buying that model or make. There is a deep satisfaction attached to working with your hands. Fine tools go right along with that feeling. Some people may argue that the less expensive tools give identical performance. If that were true, the expensive toolmakers would be out of business. When your skill level rises, the differences become more evident and hidden to the casual user. Some of the differences are admittedly small, but I think the additional cost is well justified in most cases.

I will admit the reverse is also true. A personal case in point has to do with some sheetmetal snips. For $150 I bought a pair of the most expensive European sheet metal snips with carbide cutting edges and high expectations on their performance. For that kind of money for a single pair of snips, I expected them to make lunch for me after they were done with all the cutting. On the recommendation of a real thrifty friend of mine, I also bought a pair of $20 snips of a brand I had not tried. The $20 "offshore" snips run circles around the high dollar snips. This is a case of straight hands-down performance.

Just like carpenters and their tool belts, all efficient metalworkers need to have a few crucial tools on their person almost all the time as they're moving about the shop. The class of work the shop does determines which tools are necessary. Police officers could not do their jobs very well if they had to

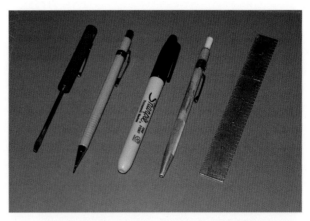

Figure 4-39: Essential tools of the trade.

make a trip back to their cars every time they needed something. The same is true for metalworkers.

It would be great if you were never more than one step from your toolbox anywhere in the shop. Everything you need would just be a short reach away. Countless hours of hunting and using the wrong tool for the job would be saved. Over the years I have tried different combinations and finally condensed it down to the bare minimum of tools I keep on my person (Figure 4-39). Having these tools at your fingertips at all times saves an incredible amount of time in a given day. You do have to develop the discipline to put them back in your pocket or they will do you no good.

Tape measure. 10-foot minimum length. You can remove the belt clip if you like to carry it in your pocket. If you can't measure at a moment's notice, you will not measure enough and more mistakes will be made as a result. A very simple way to avoid mistakes and increase efficiency is to carry a tape measure and use it often.

Calibrate your tape measure frequently. Check it against high quality rulers or combination square blades (Figure 4-40). Bend the hook to adjust the zero.

Six inch ruler. This ruler is your best friend—well maybe your second best. It will stir your coffee as well as double check a zillion things every day. Skip the really fine division rulers. There are better tools when you need to measure closer.

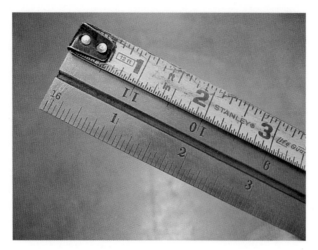

Figure 4-40: Calibrate your tape measure.

Figure 4-42: Retractable point scriber.

Know the thickness of your favorite rule. It comes in handy as a field expedient feeler gage. The one in Figure 4-41 is .015.

Retractable point scriber. The retractable point scriber (Figure 4-42) won't poke a hole in your shirt or apron pocket. A carbide tip is a must if you do any work on stainless steel.

Sharpie marker. Black or blue for light-colored materials; silver metallic for dark materials; and red for junk parts, screw-ups, corrections. I really wish I had invented this product. I have these all over the shop. I would bet that I don't have to walk more than three paces to lay my hand on one anywhere in the building.

Small flat blade screwdriver with magnet. This tool is your pry bar and pick, as well as a handy little screwdriver (Figure 4-43). I like the ones that some salesmen hand out for gifts This little screwdriver comes in handy all around the shop. I use it to align sheet metal edges prior to welding and for loosening the lock screw that holds the punch and die in my Whitney hand punch. The small magnet in the end is important also. You can use it to quickly differentiate between several types of common materials. Don't be afraid to abuse this tool. Think of it as a consumable. It will keep you from using your good tools for bad jobs.

Some kind of writing instrument for writing on paper. I like mechanical pencils because they never need to be sharpened. Use thicker .9 mm

Figure 4-41: A ruler can serve as a feeler gage.

Figure 4-43: A screwdriver, magnet, pry bar and pick — all in one!

lead for heavy handed Cro-Magnons. Good written communication is important to success. Pencils will write on vertical and overhead surfaces unlike ballpoint pens. Pencils are king out in the shop. The thin extendable tip of a mechanical pencil will reach into small template holes and otherwise go where a wood pencil can't.

Bench Work

I love working at the bench. Having all your parts and tools at your fingertips along with a steaming cup of coffee is a truly pleasurable thing. One of my most favorite jobs at the bench is to design and build a tool or small machine on the fly. I use the welding table as my chalkboard to develop ideas. Hand fitting and assembling one of your own designs and seeing it come to life on the bench is extremely satisfying.

A large portion of the work a machinist or welder does centers around a workbench of some sort. There is always something to take apart test fit or clamp down. Having a few tools at your fingertips when your working at the bench makes for efficiency.

Bench vise. Without a decent vise, a shop is not a shop. I'll tell you something: if there's any thing that needs to be high quality, this is it.

Once we were building a feed auger for a briquette press. The shop had one of those inexpensive offshore vises that I call ten footers. They look great from ten feet away, but when you get up close you can see all the Bondo used to fill the casting "Irregularities." We were using this vise to squeeze the auger flighting down onto its center shaft to get some tack welds in place. The vise failed as we worked on this screw, and just made the job that much more difficult. The boss that the handle slides through buried itself in the Bondo-reinforced front surface, knocking the clamping force in half. We responded by cranking it even tighter. Something had to give and it was the cheap vise. I was so frustrated I got on the phone and ordered a new Wilton vise for $350 on the

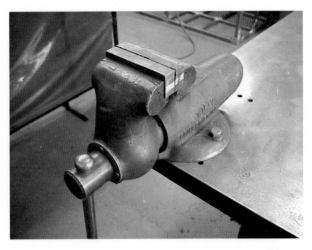

Figure 4-44: Avoid swivel bases whenever possible.

spot. I ended up in a little hot water with the boss, but in the end we still have that vise in the shop and the cheap one has probably been recycled into garden gnomes by now.

Skip the swivel base in the fabrication shop (Figure 4-44). You want the vises used in these areas to be rock solid and not move at the wrong time. Its tempting to get the swivel base. However, I guarantee it will let you down at some point when you need the vise to be solid because you can't lock it down tight enough with the little handles they put on the swivel clamps.

For most classes of work, use a maximum of four-inch jaw width. Larger than this and the vise becomes cumbersome to use. Try spinning the handle on a six- or eight-in-vise rapidly to open it while you're trying to position a red hot part. If the handle doesn't knock your teeth out, count yourself lucky.

Clamp large diameter rounds below the vise jaws. Three-point contact makes it more stable (Figure 4-45).

Install copper jaws in all the bench vises in the shop. Everybody puts protective covers over the serrated jaws anyway, so why not eliminate the marking problem? The lame flip-floppy aftermarket covers always fit poorly, which makes it a pain to grip a small delicate part. Name one thing that it would be acceptable if it has big teeth marks in it.

Figure 4-45: Three-point contact adds stability.

Figure 4-47: Mounting a vise for use with a long part.

It's a lot easier to preserve your good work than restore it. Copper jaws are easy to resurface when they get chewed up.

Copper jaws in the vise can be used as a heat sink for welding delicate parts (Figure 4-46). For the love of God, use copper instead of brass. Brass is hard and slippery and makes terrible vise jaws. Copper is soft and springy—just the qualities we want in a vise jaw.

Be sure to mount the vise so that a long part clears the bench below the jaws (Figure 4-47).

Hacksaws

Be sure to hang a hacksaw as close to the bench vise as possible. For small quick cut offs at the

bench, a hacksaw saves time over the walk to the band saw.

For right-handed people, hacksaw on the right-hand side of the vise. It will save you many sets of barked knuckles.

Keep two or more hacksaws handy with different blade pitches (Figure 4-48). One should be 32/teeth per inch and the other as coarse as you can find, like 14/teeth. This saves changing the blades, which nobody likes to do anyway.

The hacksaw will beat the cut at the power saw, including the walk to the band saw up to .75 diameter. Most guys stand at the horizontal band saw, waiting mesmerized by the slowly circulating blade for their cut to be complete anyway. So save the walk and hacksaw small stuff.

Figure 4-46: Using copper jaws as a heat sink.

Figure 4-48: Hacksaws with different blade pitches.

Figure 4-49: Your fingernail can serve as a guide fence.

Start your hacksaw cut using your fingernail as a guide fence (Figure 4-49). It gets a groove started in the exact spot and keeps the blade from jumping out. Put the keeper part of what you are hack sawing in the vise, slightly below the top of the jaws (Figure 4-50). This will keep you from spoiling the part, if you slip out of the saw groove.

Put a twang in your game and use only quality high-tension hacksaw frames. No others are even worth the energy to throw them in the scrap bin.

"I have a little sawing job for you"

When I was just a pup and first starting out in metalworking, I worked at a shop that built accessories for off-road trucks. One area in the shop did suspension and engine work on some of the vehicles.

Figure 4-50: Hack sawing in the vice.

In the other area, we did welding and fabrication, making bumpers and roll bars for jeeps and pickups of all different flavors.

At that shop I was low man on the totem pole. In fact, I couldn't even see the totem pole because I was still in the dirt with all the other worms. All the nasty, dirty, funky jobs came my way. This is part of what I call the character-building period in a person's career. Everybody has to go through it so they learn how to properly torture any newcomers. If you wimp out and quit during this period, you never get to tell how you passed on your hard-earned learning to the next generation.

I remember specifically it was full-on summertime. Inside the welding shop, it was probably 115 degrees. The summer sun beat down relentlessly on an un-insulated solar absorption roof-like covering. The concrete walls seemed to reflect the heat back into the shop to make a great oven for drying beef jerky or young shop helpers.

I worked there after school and would come in during the afternoon to work a few hours. One day I came in and there was a truck up on jack stands in the repair area with a small pool of liquid under it. I went in to see what the chief wanted me to work on.

The chief told me he had a special job that needed to get done today. He showed me the job and told me what needed to happen. He had the little grin on his face that tells you he feels sorry for you, but this is what the low man on the totem pole is really for.

The job was to cut off the knuckle joint on the front end of a truck differential. Whether you know what one of these looks like it doesn't really matter because I'm going to tell you in excruciating detail.

The cut needed to be made in the tube that connects the knuckle joint to the differential. The tube was about three-and-a-half or four inches in diameter. At the time I didn't know the wall thickness of the tube, but I was going to find out.

Somebody had already generously started the cut for me. They had managed to cut maybe a quarter-inch deep into the tube wall before they threw in the towel, hence the small pool of DNA evidence under the truck. For some reason that I still to this day do not understand, the cut had to be made while the differential was still in the truck This meant that the surgery would be performed inside the fender well of the truck.

My only weapon at my disposal was a cheap, stamped sheet metal hack saw. You know, the kind you walk past at the flea market even when the guy only wants fifty cents for it. Prisoners on death row would rather have a pack of dental floss than one of these hacksaws.

This was long before the days of sazalls. Besides, the owner was pretty cheap and had some less-than-five-dollar-an-hour labor to do the dirty work. I think this was even before decent bi-metal hacksaw blades were available for hacksaws.

So I went ahead and got cracking. Up underneath the fender of a truck, inside a tin-roofed concrete building, in the summer, with a hacksaw and a few cheap blades. I don't remember much of the next several hours other than it was taking a long time to make any progress. It sure would have been nice to have something other than a thirty-two tooth blade about as sharp as a marble to work with. The passage of time is warped in my memory. At this time in your life, if you have to do anything for more than half an hour straight, you are bored out of your skull.

The job was further complicated by the fact that you couldn't get a full run with the saw. Your available stroke was shortened to about half of normal by the cramped conditions inside the broiler fender well. So if the blade was twelve inches and the tube was four inches, that left you with way, way too short a stroke. I can't remember what the heck was in the way, but it made the job at least twice as hard as it needed to be. I'm sure I left some of my DNA inside that fender before I was done.

Saw, saw, saw the tube, slowly all the way. Sung to the rhyme of "row, row, row your boat." I sawed and sawed and sawed some more. The sweat was running off me like a garden hose run over by a soil aerator. At least the shop provided sodas in the

fridge and they were ice cold. I would quaff one in a single pass and not even taste what kind it was. Switching blades was only an exercise in futility. On the one hand, it gave you a short break from sawing. But then you realized why you had taken that blade out in the first place. These hacksaw blades were so bad that you couldn't trade them to a prison escapee for his overcooked brussel sprouts.

It turns out the tube wall was something like one-inch thick. I cannot remember the total time it took to make the cut, only somehow I made it to the end.

Amid the smoking wreckage of toothless friction blades and aluminum soda cans, I held up the life-less head of the beast slowly dripping its stinky differential lube. The big boss saw I finished with the heinous task and waddled out of his air conditioned office smoking, one of his pencil thin, little, plastic-tipped cigars. He looked down at the tube end just above the quickly evaporating pool of sweat and diet soda residue and said to me,

"Geez you sure cut that crooked. It's going to take you forever to grind that straight."

He turned on his heel in a little swirl of perfumed cigar smoke and went back into the air conditioned office. This is how I learned to appreciate a great hacksaw and premium bi-metal blades. I guess he didn't feel the wind from the knuckle joint swishing past his receding pattern balding head as he walked away.

Filing

Many people overlook this as an important skill and pick up a grinder instead. Bad filing as well as bad grinding can be seen from halfway across the shop. As my old teacher would say, "A blind man would be pleased to see it." The point is a blind man would be pleased to see anything. (Strange, he said this a lot to me. . . .)

Superb hand filing is a skill that is learned over a long period of time. One of its major advantages over powered methods is extreme control. When you cannot afford to make a mistake removing material in a delicate area, use a file. The file is directly connected to the best computer on the face of the earth. With a power tool, by the time the order to stop makes it to central command, it's often way too late. Just take a look at the work an experienced die-maker does to see what can be done with files and hand work.

When I took metal shop in high school one of the skills we had to learn was proper filing. The shop teacher was quite the thrifty little guy and would replace the files only if they made better cake frosting knives. Whenever a new file was put out in the general population, it was treated like the prom queen. And just like the prom queen everybody wanted it but only one guy got it at a time. My little trick was this: if I got the new one, I returned it to one of the extreme ends of the file rack. That way the next day I would swoop in and just grab the end file while all the others were fingering each file to find the new one they were sure was in there.

Our first filing project was a perfectly square cube. We were given a hunk of cold rolled steel that looked like somebody held a rat by the tail and gnawed it off the end of the bar. We were then required to file and sand this into a perfectly square cube of specific dimensions. If you ever need to think of a devilish little project to keep an apprentice out of your hair for say a month or so, this is the ticket. It sounds simple enough, but in practice to get all the sides to size and square is non-trivial. Try this with a file about as sharp as a putty knife and see what I mean.

Grind at least one safe edge on all your primary files (Figure 4-51). This reduces the chances of gouges and keeps you from scratching when filing up to a shoulder on your nice machined parts (Figure 4-52).

Round the noses of all your files (Figure 4-53), in particular any files used in the lathe. There is no reason to have the nose of the file square and sharp unless you scrape paint with it.

Put handles on all your files. This gives you much better leverage with the tool and makes

Figure 4-51: Grind a safe edge on your primary files.

Figure 4-52: Safe edges reduce gouges.

them much more efficient. Never use files without handles in the lathe or other power machinery! Besides, having handles lets you hang your most used files in a professional looking rack and keep them at your fingertips (Figure 4-54). I use these bright blue handles so that I can spot my files from across the shop. Many knuckleheads don't put handles on their files so that the files will lay flat in the drawers of the toolbox. Resist this tendency.

If you go in the field with a small toolkit, take at least one file. Make it the very versatile half round with a bastard or smooth cut. Half rounds fit into acute angle corners well and will dress a radius as well. It's like three files rolled into one.

Figure 4-53: Round the noses of your files.

Figure 4-54: Treat your files professionally.

If you have to file a radius smaller than the file radius, use very light pressure and twist the tool as you file.

Get rid of dull files. They actually do wear out. They can be sharpened chemically if you have sentimental attachment to them. Impress the apprentices by generously giving them your dull castoffs, "Gee, thanks mister." Resist the temptation to laugh, at least until they walk off with the misty look still in their eyes.

Grind an edge on a file for knocking berries loose after welding (Figure 4-55). The finish from knocking berries off with a file looks more professional than chasing around the entire weldment with a grinder.

Figure 4-55: Your file can be used to knock berries loose.

When filing a radius, follow the radius through the arc with the file kept in flat contact through the arc to keep it true (Figures 4-56 and 4-57).

The lines from filing follow the radius around the corner (Figure 4-58). The file needs to be kept flat against the work so the corner is not rolled off to the sides.

Can I have a new file?

My old teacher Doug controlled the hand tool inventory at the shop we worked at together. We did a fair amount of stainless, so every once in a while you would need a new file. Everybody knew what unlocked drawer they were kept in, but for some reason nobody helped themselves. In the back of your mind you knew that the old miser could remember the exact date you got your last file and

Figure 4-56: Filing a radius.

Figure 4-57: Keep the file in flat contact through the arc.

Figure 4-58: Lines from filing.

do a lightning fast cost-value analysis on your personal file consumption. If you were perceived as trying to jump the gun on a new file or, heaven forbid, hoard a new file, you would be scrutinized like Colonel Sanders trying to enter a chicken farm.

The normal drill was to go up and ask Doug for a new file when you needed one. This was no small effort. You really needed one when you worked up the courage to actually ask him to dole one out to such a pathetic excuse for a metalworker. The real frustration was he would always give you the used file off his workbench when you asked for one. It was a quicker way to get rid of you and get back to work.

Now everybody knew that a mere two feet away was the pirate's hoard of brand new files, still wrapped in the anti-rusting paper, that you really wanted. Who wants somebody's sullied

hand-me-down, used file of dubious family history? Thank you, but not me for sure.

Doug would thrust the file at you and say "Off you go." That was the end of it. No discussion, no negotiating, end of story. I would sulk back to my bench with my "new" file, truly depressed at the thought of another three months sentence to filing deprivation.

A couple of months later I happened to notice one of the other guys walking the plank to get a replacement file for himself. Suddenly I got the brainstorm to go in right after this guy and get a new one for myself. There was the potential risk of being branded a copycat, but it was worth it. I was certain to get a fresh one out of the drawer for myself, knowing the handout system so well.

I know I had that little smirk on my face when you think you're smarter than everybody else. I almost went as far as to rub my hands together in greedy anticipation. I gleefully watched the other creature crawl off with his nice used file back to his cell in the salt mine, thinking to myself "thanks for taking that bullet for me." I suppressed a little "waaa waaa" for the poor guy.

I wiped the smirk off my face and headed up to the poop deck.

"Hey Doug, can I get a new file? My old ones had it from all those stainless tanks." He hesitated and looked at me with the hard, unwavering squint of suspicion in his eye. "Didn't I just give you one a while back?" It wasn't really a question. More like a slap across the face with a sword-scarred gauntlet. As he drummed his fingers on the table, I managed to squeak an answer, "Umm, ahh. I don't think so. It's been a few months at least."

I had carefully positioned myself between Doug and the drawer with the files in it. Almost there, hold on a little longer I said to myself. "Alright, then go on then, fetch one out of the drawer there. Be quick about it. I don't want the whole shop beating a path to my door for files."

It was hard to keep the winning smirk off my face. I turned and squelched a little snicker. I opened the drawer and looked down in horror.

The box was empty.

I pawed my way around the inside of the drawer like a dog digging at a gopher hole frantically hoping to find at least one new file. When I finally came up for a breath of stale air, I said,

"Hey Doug, the box is empty." I turned around to look at him, trying to hide my utter disappointment. Doug snaps his fingers and says, "That's what I was trying to remember. Thanks for flogging my memory. Here, take this one." He reached and handed me a really sad–looking, used one from the grinding dust-caked recesses of his bench. This one's got plenty of life left in it. Off you go now."

This is how I learned to appreciate a new file. From that day on, every time I use a file I can't help but think of how a nice, new one cuts.

More Tools

Use a cordless drill to hand tap through holes (Figure 4-59). This is at least a thousand times faster than twirling by hand for a dozen holes. It beats setting up the tapping head for a few holes or when there are a few different sizes. Heck, use two drills if there is more than one size hole to tap. The block in Figures 4-59 and 4-60 is a tap guide to keep the tap perpendicular to the part. These can be made quickly for those jobs that don't lend themselves to a universal type block.

Mark the age of your re-chargeable batteries (Figure 4-61). These batteries eventually wear out.

Figure 4-60: Using a block as a tap guide.

Figure 4-61: Indicate the age of re-chargeable batteries.

If you're like us and have a bunch of them, it's hard to keep track of which ones cannot hold a charge and should be replaced. These can usually be taken back to a store that sells rechargeable batteries and returned for recycling. Typically they last for only 500 charges or so at best.

Small cantilever clamps make great tight quarters tap wrenches (Figure 4-62). The jaws are even grooved to hold a tap securely.

Figure 4-63 shows several special speed tools I have made over the years. We try to use as many of the same size fasteners as possible when we design machinery. Therefore, I made up these ball driver tee handles for the most frequently encountered sizes. The short, fat one is for removing the socket head screws that hold the jaws in a Kurt vise. The hex blades are silver soldered into the handles and can be replaced as needed.

Figure 4-59: Using a cordless drill to hand tap through holes.

Figure 4-62: Using cantilever clamps as tap wrenches.

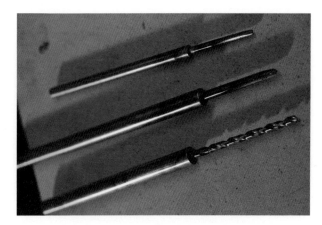

Figure 4-64: Simple tap extensions.

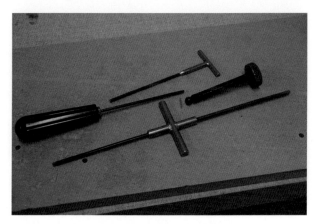

Figure 4-63: Special speed tools.

Figure 4-64 shows a couple of simple tap extensions you can make when you have to tap a long range hole. These will reach into those deep, dark recesses you have to tap a thread or just chase an existing thread. The taps are silver soldered into the extension. When they get dull, you just heat it up and pop in a fresh tap. Be sure to make the corresponding tap drill with an extension also.

Saws and Sawing

Every metal working shop usually has at least three kinds of saws. If they don't, then their process methods should be reviewed. Saws are extremely efficient at cutting almost any material. Use the efficiency example of the metal supply center or lumber processing industry. These thin kerf methods have been optimized over a long period of time; saws are a critical part of their daily operations. Many shops underutilize their sawing dollars by machining or grinding away material that could be more efficiently sawed away. Don't overlook the chunk scrap produced by sawing—it has a higher per pound value than turning or milling chips.

The factors that make sawing so efficient are a thin kerf which wastes less material and uses less horsepower to remove material. Compared to milling sawing is quite rapid. By the way, when you buy raw material from a material vendor, you pay for the saw kerf which goes in their scrap bin, but you pay good material price for it. These guys have had years to figure out all the kerfs and profit angles.

If you're just setting up a shop and are thinking about what kind of saw to get first, I would strongly suggest the vertical band saw. These highly versatile saws can cut almost anything. With a few different blades, you can cut anything from meat to hardened steel or even glass. The key factors in selecting a vertical saw are speed range and throat depth. Buy the maximum you can afford in both of these. A blade welder is nice, but not a deciding factor. Commercially welded blades are cheap in comparison to welding them in the shop. It's nice to be able to cut a blade and insert it into an internal cut, so the ability to connect a blade quickly is an excellent skill to develop.

Figure 4-65: TIG welding band saw blades with silicon bronze rod.

Figure 4-67: Exposing the weld area.

Figure 4-68: Using an abrasion device to remove buildup and level the seam.

If you don't have a blade welder or you ran out of silver solder, you can TIG weld band saw blades with silicon bronze rod (Figure 4-65). The weld is annealed a second time after grinding to remove brittleness along the heat affected zone (HAZ). Be sure to accurately line up the spine of the blade or it will make goofy clicking noises as it runs through the guides. This example is one of those special right hand left hand blades (Figure 4-66).

After you weld, silver solder, or otherwise connect the two stray ends of a band saw blade, you need to remove the weld buildup and level the seam. Use a curved surface to expose the weld area to your preferred abrasion device (Figures 4-67 and 4-68). After grinding, peen the weld area on a flat dolly to level the seam so the blade thickness is correct (Figure 4-69).

The mouse

One Monday morning we came back to work to find a mouse trapped in the shop sink. The little fuzz-bot must have been hunting for food and fell into the deep plastic sink. I don't know how long the little rodent was sitting down there, but it must have been quite a while. When we found him, he was barely hanging on. He made no effort to run when I picked him up by the tail. He flopped around rather weakly, which must have been about the point I felt a little sorry for him. It was wintertime so I figured my newfound friend must have been half frozen sitting down in that sink bottom all weekend. I put a piece of paper on the vertical band saw table and turned on the gooseneck light, positioning it close enough to warm him, but not too close to make mouse fritter. Somebody nearby

Figure 4-66: A right hand left hand blade.

Figure 4-69: Peening the weld area.

thoughtfully suggested that the mouse was probably hungry from all those hours sitting in a shop sink with nothing to eat but small metal filings. They generously coughed up a tortilla chip plucked from their lunch bag for our new friend. I broke apart the tortilla chip (nacho cheese) in to small tidbits and then was called away to do something else.

An hour or so later I realized I had forgotten the new shop mascot and went back to check on him. He was definitely doing better. He was sitting up with his eyes closed delicately holding a tiny triangle of nacho cheese flavored tortilla chip with both cute little grippers. I then got the bright idea that a little pure oxygen might do him some good. I had recently read an article about the new fad of oxygen bars and how a herd of oxygen-sniffing humanity was rapidly growing. It may seem obvious, but the article described the many benefits from breathing oxygen-enriched air. Cures for headaches and ingrown toenails were claimed along with increased vigor among other things. I figured if it was good enough for humans, it wouldn't hurt our mouse friend. I wheeled the oxy-acetylene rig over from the welding shop and turned on the oxygen to barest possible minimum. I carefully tested the pressure on the exposed skin of my cheek, much like a mother testing the temperature of a baby's bottle. I picked up the little furry patient and cupped him in my hand as I gently fed him pure restorative welding grade oxygen.

About this time the tortilla donor came by to see how the patient was doing and give me some general flak for taking in an orphan rodent. Another person came over to see what all the fuss was. I opened my hand a little to show the non-believers how well my little buddy was doing. I guess all the attention of three large primates staring at him with those big round eyes was enough for him.

He spun with vicious speed and charged the quarter inch or so to my pink fleshy thumb and bit down with his dingy buck-tooted incisors. Now most shop people have pretty tough hands, but this potentially rabid, little micro rat managed to clamp down on the tiny little callous-free cuticle of my thumb between the nail and the side meat. Needless to say I hastily ceased his oxygen therapy by dropping the torch and yelping with my hand held out for all the very entertained spectators to see. I think he couldn't let go because his nacho cheese flavored teeth were sunk into my thumb. Maybe it was all the laughing or possibly me shaking my hand vigorously that finally persuaded him to let go.

Convinced of the mouse's miraculous recovery, I picked him up carefully from his distal end and took him outside and sent him on his way.

Protect critical surfaces with tape before cutting (Figure 4-70). A good grade of masking tape works well for a cutting guide also. The clear sheet shown in Figure 4-71 is called Surface Armor and is available in many different tack strengths. This

Figure 4-70: Protecting surfaces with tape.

Figure 4-71: Surface Armor protects important surfaces.

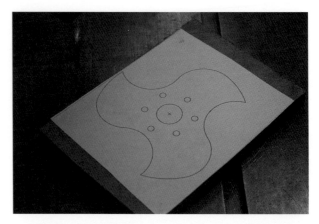

Figure 4-74: Paste a full-scale drawing to the part.

material protects your important surfaces and removes without leaving behind a residue.

Lay a couple of lanes of protective tape on the saw table to keep your nice material blemish free (Figures 4-72 and 4-73).

Paste a full-scale drawing to the part and band saw without layout (Figures 4-74 and 4-75). Use a

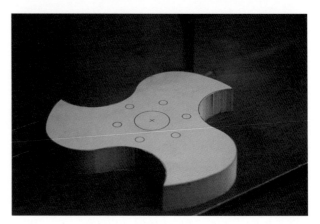

Figure 4-75: Band saw the part.

Figure 4-72: Use several lanes of protective tape.

Figure 4-73: The tape protects your material from blemishes.

little spray contact cement to attach a drawing directly to the part. This is a great trick for prototyping. Laser printers are so accurate, I use prints from them for radius gages and checking templates all the time.

Use a wood cutting circular saw to cut aluminum plate (Figures 4-76 and 4-77). Be sure to wax the blade. Use the cheapest carbide-toothed blades you can find at your local home center. High-quality blades don't seem to hold up any longer than the cheap ones. I have personally cut 2-inch thick copper with this setup. If you use a guide, you almost have a plate saw.

A wood cutting miter saw makes short work of aluminum and plastics (Figure 4-78). We added a fence with a recessed scale to keep from scraping the stick-on ruler off (Figure 4-79). Four-inch diameter aluminum is a snap with this setup. Use a carbide-toothed blade for cutting

Figure 4-76: A wood cutting circular saw.

aluminum. Wax the blade carefully for best results.

One of the great abuses of band saws is not changing the blade for cutting thicker or different alloys of material. If I had unlimited floor space and a budget to match, I would have three vertical bandsaws, each set up differently. But since I don't live on Fantasy Island, I have to change the blades along with everybody else.

I don't know why people don't get this one. The chart has been on the front of the machine as long as I can remember, but here it is condensed to:

Thin stuff = lots of smaller teeth
Thick stuff = less but bigger teeth

Take a look at the blades they use to saw trees into lumber for a clue. These have about a 4- inch pitch between each tooth and the gullet is big

Figure 4-77: Using the saw to cut aluminum plate.

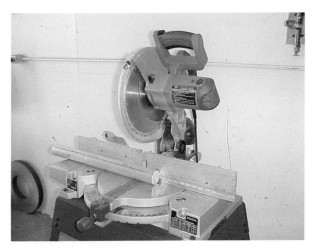

Figure 4-78: A wood cutting miter saw.

Figure 4-79: Adding a fence with a recessed scale.

enough to park a chip the size of a cinnamon roll in it.

For most practical purposes, the average shop can get along with two blade pitches. This will help cut down on the confusion for people that don't look at the blade chart.

Use your old wasted blades for friction cutting sheetmetal (Figure 4-80). Run the machine at the highest possible speed and wear ear protection. This makes short work of cutting sheetmetal of every flavor up to about 10 gauge for average horsepower machines. This also works for cutting hard stuff like linear bearing shafting. The process is not anything like sawing because the material is heated and softened to its melting point and carried away by the moving blade. You

Figure 4-80: Friction cutting with old wasted blades.

Figure 4-82: Cutting square corners in a tight spot.

can see in Figure 4-81 the heat discolored burr in 10 gauge stainless steel. The process is limited by available horsepower when you get into the thicker materials. The main trick is using a high SFM cutting speed, so run it as fast as your saw will permit.

When cutting square corners in a tight spot, use the trick demonstrated in Figure 4-82, instead of trying to steer a blade that's too wide to make the required turn. Blades that are 1/2 wide are a good general purpose choice and cut very straight in the straight sections (Figure 4-83), hold up for long periods of time, and are not as sensitive to blade tension issues as narrow blades. If you want to make a good compromise and pick one size, then make it 1/2 . The tradeoff is less turning ability.

Start your cut and just steer for the opposite corner (Figure 4-84). Come up right into the corner to establish its position, then back out. Do the same for the opposite corner (Figure 4-85). Alternate this way until you can actually start the blade in the short direction. Use the front of the blade and work sideways like a die filer to clean out small features. Use a good fresh blade for best results. Band saws are superb roughing tools. Who needs a milling machine anyway when you have a good band saw (Figure 4-86)?

Cut large block-shaped parts out of large rounds (Figure 4-87). In some types of material, large blocks are not readily available. You can rough saw them out of large rounds, which are sometimes the only form you can get on short notice.

Figure 4-81: A heat discolored burr.

Figure 4-83: Cutting the straight sections.

Figure 4-84: Steer for the opposite corner.

Figure 4-86: A good band saw works well.

Three dimensional sawing can save huge amounts of time in the machine shop. Be sure to leave enough stock for cleanup cuts when you use this method. Better to take a couple of extra passes than have a spot that has raw saw marks in it. In this example, I have laser printed the drawing full scale and bonded it to the raw material with spray contact cement (Figure 4-88). Any holes or corner reliefs are added next (Figure 4-89).

Cut the features first that allow you to keep the blank as block-like as possible (Figures 4-90 and 4-91). The danger is that you cut away your guide lines or end up with something that's difficult to manipulate in the saw. The vertical band saw is one of the most efficient metal removal tools in the average shop.

Forget one of those wimpy pathetic television machines for working out your abdominal muscles. Make a belly board and cut some 1-inch stainless steel for a few hours (Figures 4-92 and 4-93). Six pack abs here I come, or maybe it's just a six pack here I come. . . .

Get used to using a push stick to sweep the scraps off the saw table. Develop the habit of sweeping from behind the blade toward yourself and the operator side of the saw (Figure 4-94). If you were to accidentally hit the blade, the last time I checked there were no teeth on the back of the blade. You will get a good scare but still have all your fingers intact.

Check out the push stick in the picture (Figure 4-95). I wonder how it got so many saw cuts in it. You think it is an effective tool?

Figure 4-85: Repeat for the opposite corner.

Figure 4-87: Use large rounds to cut large block-shaped parts.

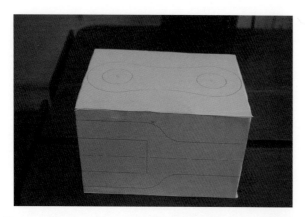
Figure 4-88: Bond laser printed drawing to the raw material.

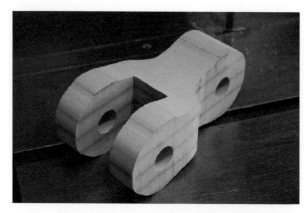
Figure 4-91: Continue shaping the part.

Figure 4-96 shows a great little vertical band-saw vise we found in the last year or so. You can clamp those pesky little jobs where you are more worried about clipping a finger than doing a good sawing job. Short pieces of round bars are easily clamped for a spin-less stress-free cutoff (Figure 4-97).

Grind the weld edge preparation with the teeth facing opposite directions (Figure 4-98). Twist the blade 180 degrees and then grind the ends square, prior to welding. The grind angles will match perfectly for a perfect weld (Figure 4-99). After grinding, I like to add another quick anneal to the weld area to remove any brittleness that may have

Figure 4-89: Add holes and corner reliefs.

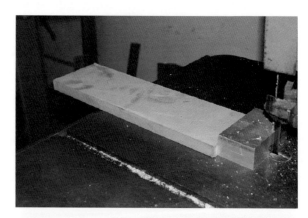
Figure 4-92: Making a belly board!

Figure 4-90: Cut features that keep the blank block-like.

Figure 4-93: Cutting 1-inch stainless steel.

Figure 4-94: Sweep from behind the blade toward yourself.

Figure 4-97: Clamp short pieces of round bars.

occurred during the weld finishing (Figure 4-100). I used a crummy old friction blade for the photos, so don't wonder why I'm welding a trashed blade. We don't weld many blades in house anymore. The prices for welded blades are pretty much in line with blade coil prices so we save the labor and let somebody else do it.

If you have a bunch of small rods to cut in the horizontal saw, drop them in a channel and weld the ends together (Figures 4-101 and 4-102). This will keep any of the rods from rotating or changing lengths while you do multiple cutoffs (Figure 4-103). By grouping small sizes together like this, you can now take advantage of the free cutting abilities of coarser pitch blades to make short work of a large number of pieces (Figure 4-104).

Weld your short stump of material to another piece to get another couple of cuts in the

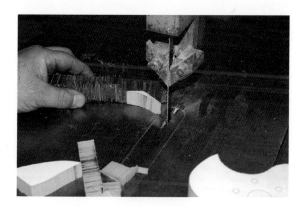

Figure 4-95: A push stick protects fingers.

Figure 4-98: Grinding the weld edge preparation.

Figure 4-96: A vertical band saw vise.

Figure 4-99: Preparing for a perfect weld.

Figure 4-100: Adding a quick anneal to remove brittleness.

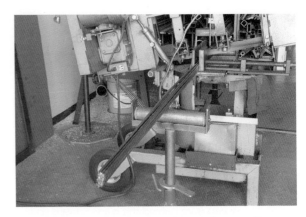

Figure 4-103: Preparing for smaller cutoffs.

saw (Figures 4-105 and 4-106). Or you can use a jaw spacer to keep the vise jaw from tilting (Figure 4-107).

When making multiple cuts on angle, stack the pieces so the vise pressure clamps the lengths together (Figure 4-108). Three-point contact makes for firm clamping.

When you cut miters in the saw for weld prep, cut the angle just a fraction of a degree more or less than 45 (Figures 4-109 and 4-110). What this does is make sure the outside or the inside of the joint is in contact. When you tack weld the corners that are in contact (Figure 4-111), the frame is still free to move around a bit to get it squared accurately

Figure 4-101: Cutting small rods in the horizontal saw.

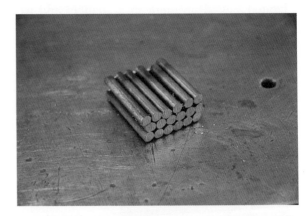

Figure 4-104: Coarser pitch blades shorten the work.

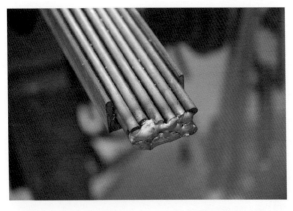

Figure 4-102: Weld the ends of small rods together.

Figure 4-105: Weld a short stump to another piece.

Figure 4-106: Making additional cuts in short stumps.

Figure 4-109: Cutting miters in the saw.

(Figure 4-112). If you cut the parts so the inside of the frame touches first, it can save joint preparation time on the outside of the miters. Pay attention to which side you tack first. When squaring you want the tack to act like a little hinge.

Sometimes its easier to cut a cross member a little short instead of grinding a big weld preparation

(Figure 4-113). This also pays dividends when finish grinding because the finished weld is closer to flush with the surface (Figure 4-114).

When you need to do a really accurate job on the vertical band saw, try using two layout lines with a spacing as close as possible to the blade size (Figures 4-115 and 4-116). When you follow

Figure 4-107: Using a jaw spacer.

Figure 4-110: Use a degree more or less than 45.

Figure 4-108: Preparing for multiple cuts on angle.

Figure 4-111: Tack welding corners that are in contact.

Figure 4-112: Squaring a corner accurately.

Figure 4-113: Cutting a cross member a little short.

Figure 4-115: Using two layout lines with a vertical band saw.

Figure 4-116: The spacing is close to the blade size.

a single scribe line, it's easy to wander double the width of the blade and still stay on a single scribe line. Having two lines is like driving a car between two telephone poles and not clipping your mirrors. This trick works well for splitting a two-piece part that the designer has not left enough sawing and cleanup allowance.

For all the designers out there, a minimum starting number for splitting and finishing a two-piece part should be 1/8 in easy materials and moderate thickness up to 1.00 or so (Figure 4-117). The thicker the stock, the wider the split needs to be. There are two ways to do this type of operation quickly and cleanly in the average shop. In most cases, you can use a slitting saw and slice right

Figure 4-114: Finish grinding the cross member.

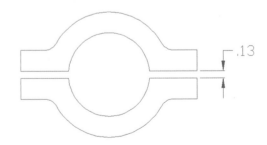

Figure 4-117: Splitting and finishing a two-piece part.

through without the need to re-surface the cut, or you can band saw and re-surface the cut face. The problems come up when there is too little allowance for cleanup. Thicker slitting saws are more stable in deep cuts and can be fed more efficiently than skinny wobbly large diameter saws.

One trick we use for creating a split part with a zero gap out of one piece of material is to start out with a diameter slightly larger than the finish size, and splitting in half first (Figures 4-118 and 4-119). In many cases, the designer needs a part that has the split exactly on the centerline, which makes for twice the work and twice the material in the shop. With expensive material or large sizes, this can make for a big cost increase.

This pre-splitting trick works well if the part has any features like screws or pins to locate the two

Figure 4-120: Using screws and pins to fit two halves together.

Figure 4-118: Creating a split part with a zero gap.

Figure 4-119: Start with a diameter slightly larger than the finish size.

halves together for the finish tuning and boring, like this Tantalum part seen in Figure 4-120. A bar a bit larger than the finish diameter was split and then held together to create this centerline split. In this case, wire EDM was used to split the part to minimize material loss. In some cases, the parts can be also be tack welded together if they don't have any fasteners, with the last operation being to remove the small welds. This saves making two parts to get one complete assembly with an exact centerline split.

"Can I borrow a hole saw?"

One of the guys I worked with at a sheet metal shop had a little home project he was working on. This guy was installing a set of stereo speakers in his truck. Normally this is a pretty straightforward job, except this particular guy had less than perfect luck. Have you ever known anybody who when something bad happened, it always happened to him? Well, this is the guy.

The shop we worked at had a unique tool lending policy. You could borrow small hand and power tools with permission, of course, to take home and use on personal projects. The only rule was you had to bring it back every day, even if you weren't finished with it. This was in case there was a company job that required the tool. If you screwed up and forgot the tool, you had to go home on your own time and get it that day. This was a pretty good incentive to do the right thing and always bring the tool with you, even though you would just haul it home again and again until you were done. Incidentally, this is the same shop where I was enlightened to the Thursday night policy.

One day my good buddy Steve asked if he could borrow a four-inch hole saw and the hole hog drill. If you don't know what a hole hog is, it's one of those high torque drills with two speed ranges—slow and glacial. On the high speed, this drill has enough grunt to run an oil drilling derrick drilling for crude oil. On the low speed, you have enough gear reduction to beach a battleship up a boat ramp. Quite a number of people have gotten themselves into trouble with this tool because of the torque reaction. Nothing like a good arm winding like a Gumby toy to start the day.

I saw Steve carrying the drill out to his car one day after work and asked him what he's cutting. He told me about the great speakers he got and how he was going to mount them in the doors of his truck. At the time, I really didn't give it much thought. But, as an afterthought and having used the Gumby drill myself, I would have chosen a saber saw.

The next day rolled around and no Steve. I never really noticed until I saw the foreman talking on the phone and laughing afterward about something. Around lunch time, somebody asked what happened to Steve and the foreman gives us the scoop.

If you have ever used one of these drills, you know they have an auxiliary side handle made out of three-quarter-inch steel pipe. Anything that comes with a handle like that is bound to be trouble. This sturdy looking pipe is to give you the illusion of some kind of control over the tool. Its real function is so you twist your arms equally when the inevitable grab happens. Every time you use one of these, it's like tickling the tail of the dragon.

The foreman finally told us the story of what happened to our comrade. Apparently he was using this

Figure 4-121: Don't put the book down now!

drill to hole saw these speaker holes in the door of his pickup. He was sitting or kneeling on the ground to get a good grip on the drill—in case it grabbed, I guess. Well, when the hole saw broke through, it did what they always do on sheetmetal; it broke through in one little spot. Several milliseconds later the whole operation went from sub-critical to full implosion. The drill caught and spun Steve up in a nice ball of dirty laundry. At the same time, the pipe handle managed to clock him in the mouth and knock out a couple of teeth.

This all happened with him wedged between the door and the rocker panel of the truck. He had nowhere to run to get away from the flailing arms of the death spiraling drill motor, so he took in on the chin like a real trooper.

When he came back into work his face was bruised and he had a couple of butterfly bandages on his nose. I never heard how the speakers sounded. . . .

In this last section, we discussed quite a few things to get your shop up and running efficiently. Don't make the mistake of underestimating the importance of the basic setup and infrastructure items in your own shop. If you are working on reducing overhead and increasing profit and efficiency, you sometimes have to get into the small details of how things get done out on the shop floor. Good leaders and owners should question their methods from time to time and ask the question, "Is there a better way?"

Rigging and Lifting

"Humans were invented by water to carry it uphill."

This section deals with lifting, moving, and handling materials and machinery in and around the fabrication and machine shops. The ability to safely and quickly handle and secure loads is an integral part of the entire job process; it may be necessary to handle these item multiple times during construction. The more efficient your handling, the faster the job will completed.

I have grouped all these topics under the heading of rigging and lifting. Rigging deals with setup and equipment, leading up to the lifting event. In the olden days, riggers were the guys on the ground placing slings and chains while the crane operator waited for their signals to begin lifting. In modern times, rigging in the small shop falls to the guys standing closest to the equipment when something needs to be moved.

Humans are basically pretty weak animals. Without our tools and machines, we are just hairless primates with no particular strengths other than our big brains. Rigging and lifting in the shop requires brains. Cranes and forklifts seem like pretty simple machines that anybody should be able to use without any difficulty. This is pretty far from the truth. In my years of experience, it seems like the simplest machines and processes cause the most

trouble. The ease that you can move an object that weighs several tons, with no sensation of weight or even effort, disguises the seriousness of these operations. They are taken for granted and that is when the trouble starts.

If you are lucky enough to have an overhead crane or forklift, be sure that everyone who uses it has been certified in its use or had some instruction from a professional rigger. There are many subtleties and serious consequences to moving heavy things with small brains. The saying goes, "The rigger must not guess, he must know." Know the weight, know the capacity, retain the load, and control the load. These are the key elements for success in moving heavy objects. I think everybody has heard the statement, "Bend your knees when you lift something heavy." The same consideration should be given when preparing to move something too heavy to lift personally. Think before you lift and protect yourself at all times.

Whenever we handle, move, or load equipment or materials, it is the responsibility of the rigger to make sure the entire process is carried out in a safe manner for the next person in the job process. This includes landing the load so it is safe, as well as securing the load for safe transport or, at a minimum, supervising the work to ensure proper execution by others. Typically the rigger has overall responsibility during a lift or move. It's important to note that only one person should give instructions where multiple people are involved.

Rope Work

Everybody should know how to tie at least two knots. My two favorites that can be used almost anywhere for multiple purposes are the good old Bowline (Figure 4-122) and a simplified version of the trucker's hitch (Figure 4-123). The first problem that seems to plague people when they pick up a length of line is, "I need a loop." The second problem is, "I need to get the slack out of this line and get it tighter than all get out." Everything else falls in line behind these two needs. There are hundreds of useful knots, but what I tell anybody who asks me is just learn two. If you remember and

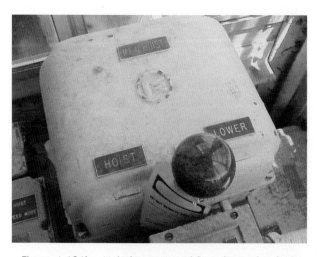

Figure 4-121b: Main host control for a large dry dock crane at Mare Island Naval shipyard.

Figure 4-122: The bowline.

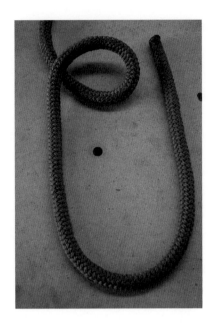

Figure 4-124: Starting the bowline.

master these two, then you most likely won't need any more.

The bowline is pretty simple and a great way to get a loop in the end of a line. They will untie easily even after subjecting them to heavy loads. I used to sail on a large sailboat and the jib sheets where attached to the clew of the jib with a simple bowline. The sheet was 5/8 diameter Dacron line. Keep in mind that this is how the wind connects to the boat and drives it through the water. I used to stand on these sheets sometime just for fun. I

weigh over 200 lbs and my weight would barely deflect the line; it had so much tension in it. It was more like a steel rod than a piece of rope. After docking the boat, the jib sheet bowlines could be easily untied in two or three seconds even after that severe loading. This is the champion of knots.

The bowline is a simple enough knot that there should be no excuses why everybody can't tie one. It's the old story about the rabbit coming out of the hole (Figure 4-124) and going around the tree and back into his hole (Figure 4-125).

Figure 4-123: The trucker's hitch.

Figure 4-125: Completing the bowline.

Figure 4-126: Starting the trucker's hitch.

Figure 4-128: Form a loop.

The second most useful knot that I have come to use over and over again is the trucker's hitch. There are several variations of this knot that work but are not easily untied or changed quickly. The beauty of this particular method is you tie no semi-permanent loops in the line. The loop is secured by a half hitch and just falls apart when you untie the load. Start out by passing the line from the load and securing it around the hook or load eye (Figure 4-126). I'm right handed so I go from left to right as shown. Cross the free end over, as shown in (Figure 4-127) while holding the bit from the load.

Pull up some slack from the free end to form a loop as shown in (Figure 4-128). This will become your "pulley" which will tension the load line.

Twist the load line toward yourself and form an eye (Figure 4-129). The loop in my left hand

(shown in Figure 4-128) is inserted into this eye (Figure 4-130).

Figure 4-131 shows the simplified trucker's hitch tied properly. As you pull on the free end, the half hitch eye tightens around the loop and forms the pulley used to tension the load (Figure 4-132). Once the load is tightened, the free end can be secured around the fixed end loop with a couple of half hitches. If you really need more tension, you can double this trucker's hitch up to get twice the tensioning power. The double trucker's gets a little messy for concise pictures, but try it once you have the single trucker's hitch down pat.

Figure 4-129: Form an eye.

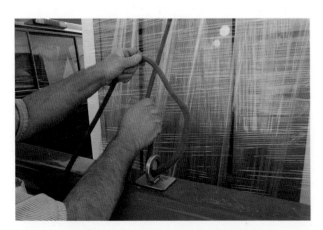

Figure 4-127: Cross the free end over.

Figure 4-130: Insert the loop into the eye.

More Lifting

Whenever you have to move a piece of material or heavy equipment with the forklift that has a steel or metal bottom, always put a piece of wood between the fork blades and the object (Figure 4-133). All you need is a thin piece of plywood which acts like a brake shoe and keeps the load in position. The oily bottoms of machine tools are a good example of metal-on-metal as a low friction bearing surface. Whenever you have a metal-on-metal moving scenario, this can mean the difference between the load shifting dangerously and a simple lift. I'm sure any of the readers that have used a forklift has experienced a long piece of bar stock pivoting as if on bearings on the fork blades as a small turn was made. Uncontrolled loads are just that—uncontrolled.

Figure 4-133: Place wood before moving a heavy object.

Figure 4-134 shows a fast simple easy to make a lifting lug without a hole. It's formed from stock flat bar sizes with two equal angle bends. Please note this type of lifting lug is not suitable for turning loads over; it is used for pure vertical lifting like you would find in tanks and boxes. It is super easy to make and add on where you need it during fabrication (Figure 4-135). These lugs work equally well with wire or rope slings (Figure 4-136). For loads that need to be turned over, you will need a weld on a lifting eye with a closed hole.

How to secure sheetmetal or other flat sheet material in a pickup truck. The best method of securing sheet material in the truck is to strap it to its shipping pallet or skid. Many times this is not possible and individual sheets or plates need to be transported. One trick we use is to secure the sheet with clamps, then use the clamps as tie points for

Figure 4-131: The simplified trucker's hitch.

Figure 4-132: Forming the pulley.

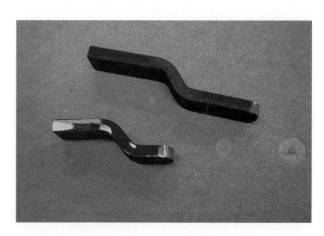

Figure 4-134: A simple lifting lug.

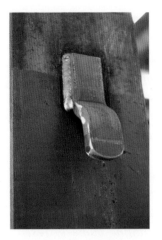

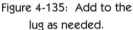

Figure 4-135: Add to the lug as needed.

Figure 4-136: Works with wire or rope slings.

Figure 4-138: Use wood backing for thinner material.

the load (Figure 4-137). For thinner material that might be damaged by the clamps, a wood backing board can be clamped to the sheet for greater security (Figure 4-138).

"Magic Carpet Ride"

This is a story with a happy ending that could have easily gone the other way. For the people involved, the planets aligned and everything came out alright with nobody losing their head—literally.

The shop I was working in at the time was extremely busy and on a crushing deadline to complete a large machine. I worked in the welding and fabrication shop and was working on a large

Figure 4-137: Securing sheet metal with clamps.

stainless fabrication made from 1/8 sheet. To our dismay we realized we were going to run short of 11 gage stainless steel sheetmetal on this critical part. I told the engineer on the project, who then mobilized the purchasing agent. After several phone calls and some mad scrambling, they found a sheet locally that we could drive over and pick up. Our shop helper was an older retired man who worked in our shop to stay busy. We cornered him and told him we needed this sheet picked up right away. If I recall this was well before lunch, so we sent the driver on his merry way.

The shop truck was a smaller foreign pickup equipped with an overhead lumber rack. For most of the material it was used for, it worked out fine. Nobody stopped to think how the driver was going to secure a 4-x-10-foot sheet of 11 gage sheetmetal and bring it back safely. Oops.

We were all working away furiously for hours before somebody realized the helper had not come back. This was back in the days before cell phones so we had no indication anything was wrong. Some more time passed before we started to really wonder what was going on. The buyer called the metal supplier and confirmed that our man had picked up the material several hours ago. Several hours? Where the heck was he? We guessed that he may have had car trouble and was sitting on the side of the road with a flat or something, or maybe he stopped for lunch. Everybody was pretty upset and impatient that he hadn't come back yet.

Finally in the late afternoon he wheeled into the yard, ashen faced and wide eyed. To his credit, he had the sheet at least. The first thing I noticed was that the haphazardly-tied sheet of stainless was bent and obviously damaged. After we got the full story, we were also wide eyed. As he told the story, we realized it could easily have been a horror show.

Everything was fine until they went to load the sheet in the mini-pickup. A 4-foot sheet doesn't fit in the bed of a little mini-pickup, so they decided to put it up on the lumber rack. From a weight perspective, this was probably alright, but once you factor in the aerodynamic properties of a sheet of 1/8 thick stainless, the idea loses some appeal. Apply a liberal dose of bad judgment and junior knotsmanship, and you have just set the stage for a spectacular event.

The driver and whoever was un-helping him at the metal supplier removed the sheet from the skid. This was the first critical mistake. I guess it fit on the lumber rack better without the additional complication of a wooden skid. At this point, the sheet was lashed down to the rack with that black and orange stripe plastic tie down rope everybody seems to use on the highway. The rope was just passed around the edge of the sheet and down to the tie downs on the rack. Did I mention that this was sheetmetal with a sheared edge? No padding or abrasion-resistant material was used between the rope and the sheet. Second critical mistake.

Combine the first two critical mistakes with the need for speed and you have the makings of a real disaster. In true form, after loading the sheet and "securing" it, the driver jumps on the freeway to get back to the shop quickly. Oops. Third and final mistake.

Partway on the return trip, the vibration from the flailing stainless steel para-wing strapped to the lumber rack sawed through the plastic retaining rope and liberated itself skyward. Keep in mind this was on the freeway at something like freeway speeds. Just to get the facts correct: the sheet weighs a little over 200lbs and started its flight

from five or six feet off the ground at 60mph launch speed. I almost forgot the other thing I noticed about the sheet when it finally came back to the shop. It had skid marks from rubber tires on it.

The driver said it got real quiet when the sheet broke loose. He looked in his review mirror and saw all the cars behind him on the freeway changing lanes at the same time. How nobody was killed or injured is still a mystery. Apparently several cars hit or ran over the sheet but were lucky enough not to be chopped and channeled before the corrosion-resistant guillotine came to rest in the middle of the freeway.

Unbelievably, he got some help from these lucky motorists to get the sheet back up on the rack. I'll give you one guess what he tied it down with. Yes, he used the same rope and the only change was he had to tie the sawed-in half ends back together so it would reach.

The metal gods must have been smiling on him for the second half of his adventure. He made it back to the shop without losing the sheet again. When we got the full story, it turns out the sheet became airborne on the freeway and at least three cars hit and ran over the sheet. Nobody was hurt but it took some time to exchange insurance information on the side of the road. I never quite heard how he got the sheet back onto the rack of the truck.

We had that bent and skid-marked sheet behind the metal rack for years after this happened. Finally one day after finding no suitable sized pieces in the rack, I ran over it with the forklift to flatten it enough to get it under the shear hold downs. I got a couple of usable pieces out of it and tossed the skid marks in the scrap barrel.

The really ironic part was it turned out we had enough material to do the original job; we really didn't need the sheet the helper went to pick up. I often think about that sheet flying through the air when I get behind a truck with something not very well tied down.

Ever change lanes with a shiver running down your spine. . . .

Figure 4-139: Three-point leveling and moving.

Figure 4-141: Using an overhead crane.

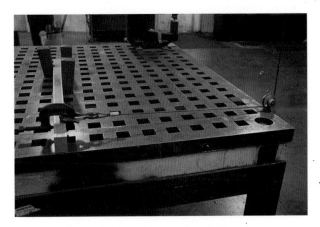

Figure 4-142: Using horizontal forces.

Three-point leveling and moving (Figure 4-139) is superior to four-point leveling in every way. It's easier and faster to level something with three contact points compared to four. Three points are like a stable tripod and are always in contact with the floor or table no matter how uneven it is. Four points are subject to tilting or rocking, which, when leveling, is confusing. Moving a heavy load or working with a rough floor can make a simple leveling or moving job a real headache, trying to balance on four points. In Figure 4-140, I am using two fingers to move a heavy table set on three points.

Using a crane to aid fabrication. In Figure 4-141, I am using the pulling power of the overhead crane through a lever to clamp where a clamp will not fit. Tremendous forces can be generated

with the simple leverage principle. If a pulley is used, horizontal forces can be used for bending, tensioning, and straightening (Figure 4-142).

I call the little devices in Figure 4-143 skateboard dollies. They're made from common radial ball bearings and a little steel in a couple of hours. These handy moving tools make short work of heavy objects. Years ago I saw a millwright use a set similar to these to move several machine tools out of a tight spot. I was so impressed I had to have a set of my own. Used on relatively smooth concrete floors, one person can move several tons of machinery with a lightweight pry bar or a gentle push. Be sure to slip a piece of wood between your machinery and the dolly. If you use three of these under your load, they will always stay in place. When using four dollies, be sure to check their

Figure 4-140: Easily moving a heavy table set.

Figure 4-143: Skateboard dollies.

Figure 4-145: Various wedges.

placement as you move the load. Uneven floors and four contact points can cause tilting where the weight is completely removed from one corner, allowing the dolly to slip out or under the load. The round bar sticking out in Figure 4-144 is to turn and steer the dollies with a length of pipe.

Wedges. Wedges have the ability to slip into places other kinds of tools won't. Figure 4-145 shows several types and sizes that are handy around any metalworking shop. The humble wedge can generate forces beyond belief when driven with a hammer (Figures 4-148 and 4-149). The thin tip of a pipe flange wedge can open enough of a gap that a pry bar can then be inserted (Figure 4-146). You can also stack steel setup wedges back to back to make quick height adjustments or create a parallel lift or wedging action (Figure 4-147).

Figure 4-148 and 4-149 illustrate just how much force can be generated with wedges. Two steel wedges back to back and a few taps with a hammer easily bow the thick steel base plate. Wedges can be used to create gaps or openings or keep them from closing down during welding.

The compact hydraulic rams seen in Figure 4-150 pack a lot of power in a small package. Extensions

Figure 4-146: A pipe flange wedge.

Figure 4-144: Turning and steering the dollies.

Figure 4-147: Stacking steel setup wedges.

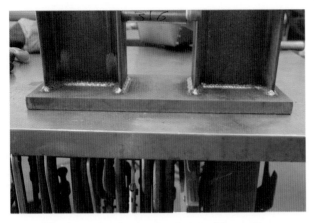

Figure 4-148: Generating force with wedges.

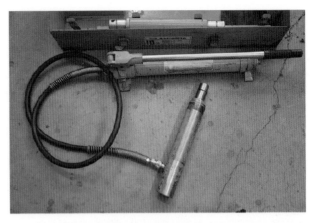

Figure 4-150: Compact hydraulic rams.

and a variety of end fittings make these useful in the fabrication shop. With 10,000 psi hand pumps, huge forces can be generated without your realizing it. A useful addition for hydraulic straightening work is a pressure gauge so you can have some feedback as to how much force you are applying. Always use extensions that are positively secured into the unit. This is another place where a thin piece of wood can keep the oily base from sliding out of position unexpectedly (Figure 4-151). I once saw somebody's front teeth get knocked out from a chunk of wood they were using as a makeshift extension. He was asked to leave the woodchuck club. . . .

Prying and levers. Figure 4-152 shows several types of basic pry bars. You will need a couple of different tip configurations for the typical work in welding and machine shops. A little block of wood or sliver of plywood keeps the bar from slipping during

a heavy pry (Figure 4-153). My all-time favorites are the screwdriver handle type. I seem to find a million uses for these almost unbreakable prying tools.

Here is a method for weighing a large object that won't fit on the scale or is outside the range of your weighing device. Hopefully you have a relatively level area where you can do this. Simply

Figure 4-151: Wood keeps the oily base from sliding.

Figure 4-149: Wedges generate considerable force.

Figure 4-152: Basic pry bars.

Figure 4-153: Keeping the bar from slipping.

Figure 4-155: Then weigh the other end.

weigh one end at a time (Figures 4-154 and 4-155). Then add the two weights to find the total. Be sure the opposite end is the only thing touching the ground. If your object must be kept level you can put a block the same height as the scale under the opposite end. The sum of the weights of the ends is the total weight of the object. Don't believe me? Stand on two scales with one foot on each scale and add the two weights.

Bucket of bolts

My old toolmaker friend told me this entertaining story about an uplifting experience he had in the naval shipyard in Washington during WWII. He was on a fairly large lathe, I gather, because the

Figure 4-154: Weigh one end of a heavy object.

shaft he was working on needed to be loaded and unloaded by an overhead crane and it took the better part of his shift to make a few roughing passes. He never mentioned what size or make the lathe was, but he sure remembered the funny part of the story.

He had finished doing the lathe work on this particular part so he whistled for the overhead operator to slew the crane over the lathe. These were large overhead cranes with dedicated operators in little booths mounted to the underside of the crane carriage. During the war, many women were working in the shipyards doing many of the jobs necessary to keep American industry cranking and the enemy croaking. This particular crane was operated by a new hire and a woman to boot. Now the story is no better or worse because a woman was running the crane; I am only repeating the story as it was told to me.

Typically they gave a 15-minute lesson on how to run the crane. Then you were on your own. I can just see how a lesson might go,

"Okay, you see this lever? This is up and down. And this one over here is longways and this one is the shortways. Got it? Oh, one more thing. If somebody yells real loud when you move one of these levers here, just let go of everything and look down."

Charlie told me that most of the crane operators sat up in the cab all day reading pulp novels and sipping who knows what out of their thermoses. The modern equivalent would be surfing the Internet and shopping for baseball cards on Ebay.

Anyways, he whistled for the crane to come over and lift this shaft he was working on out of the lathe. So far, so good. For the folks that have not worked with large overhead cranes, there are a very specific set of hand signals used to communicate your needs with the crane operator. In the case of large shipyard cranes, they could be 40–60 feet overhead. Some of the slings that are used on these big cranes are so heavy that you actually have to use the crane to go get the sling.

My buddy signaled for the operator to lower the hook so he could get the sling around the part and clipped into the crane hook. This was communicated with a signal of a specific non-middle finger pointing down and making a repeated circle with said finger. A suddenly clenched fist meant all stop.

I think he told me this was the 40- or 50-ton top main crane in the machine shop (Figure 4-156). This is the crane that sits at the top of the shop building with the most clearance under the hook. In large shops typically, they have two or even three levels of cranes with different capacities. Many times these main cranes have an auxiliary hook with a lower tonnage rating for minor lifting. In this case, the operator lowered the big mongo hook down to make this particular lift. If you have seen the hook for a 50-ton crane close up, you understand some of the forces involved in making heavy lifts. You can get into trouble real quick with one of these.

So my friend wrestled the sling around the shaft and managed to get it around the crane hook. The general plan at this point, at least in his mind, was that the crane would take the slack out of the sling so he could back out the tailstock center, then lift the shaft carefully clear of the lathe. No problem. His main concern is that he had the sling positioned properly so the shaft would sit level once the crane took the weight of the shaft. With one hand

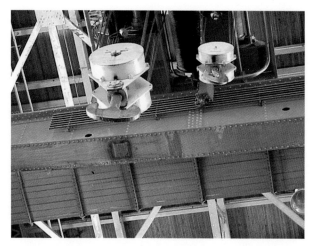

Figure 4-156: This figure shows a 40-ton main and 25-ton auxiliary hoist in the Mare Island Naval Shipyard, machine shop building 680. Notice the old-style, fish-belly riveted girders. This is the gopher's eye view someone on the ground has under one of these stump pulling cranes. These will definitely pull the bone out of meat. The reeving cables are about as big around as your wrist. On this crane, the operator sat in a sweaty little cab fifty feet off the floor.

on the shaft, my friend gave the take up signal to the crane operator. . . .

The next thing he saw was the underside of the lathe. The sound he just heard was the captain calling for the rear submarine door to be slammed shut. The entire lathe—shaft included—was hanging by the hook of the crane six feet in the air, gently swaying in utter and complete silence. Oops!

The crane operator gave full hoist upsy-daisy power and ripped the lathe out of its floor mounting, severing the electrical connection to the motor on the way up. To his credit, my friend calmly stepped back and gave the signal to lower the load to the floor—slowwwly.

I've lifted a few lathes and I would have to say that somebody was smiling on them that day. It's pretty hard to get the center of gravity nailed right out of the gate when you lift a lathe. Not to mention the live center and tailstock taking the entire load of the lathe on a teensy-weensy little point.

So, after they got the machine back on the floor, the panic-stricken crane operator climbed down and talked to my buddy. She was beside herself for providing such a spectacular spectacle and afraid she would lose her job. True to his nature, my friend calmed her down and suggested that instead of sitting up in the cab reading books, she might do a little practice on her own.

She took the advice to heart and started practicing. She eventually was able to swoop down on a bucket of bolts and scrap metal, with the bail pointing up, with three crane axes moving simultaneously. The bucket would leave the floor without spilling a single piece and be deposited on the other end of the shop, all in one smooth continuous motion. My friend mentioned that many of the other variety crane operators only operated one axis at a time: Move bridge until aligned, move trolley until aligned, lower hook to load.

As always, practice of ones trade provides precision and confident execution.

Wire, braided rope, and webbing make the most effective choker slings (Figure 4-158). When you don't want anything to slip, use a choker hitch. Unlike chain, they grip the part in a viselike noose that tightens as the load increases. Nylon and rope are used when a non-conductive sling is needed or

Figure 4-158: A choker sling.

where the load might be damaged by a metallic sling. With that said, chain slings seem to last longer and be less prone to damage than wire slings under heavy duty use (Figure 4-159). The legs of chain slings with grab hooks can be easily shortened, making them a little handier than wire for general purpose operations.

Figure 4-157: There is no shaft in this lathe, but you might need a crane to install the tool bit. I'd really hate to have this machine hanging over my head.

Figure 4-159: A chain sling.

Figure 4-160: An adjustable length sling.

Figure 4-162: Marking plumb lines.

Many times when rigging an odd piece of equipment for lifting, the combination of suitable slings is not available. You can make an adjustable length sling by using a come-along of burly-enough capacity as one of the legs (Figure 4-160). This works well with long loads with the center of gravity closer to one end.

In Figures 4-161 and 4-162, I am locating the center of gravity of an odd-shaped plate by suspending it from the crane. First hang the part from any convenient axis. After marking a line from the hanging point and vertically plumb, rotate the part approximately ninety degrees and mark a second plumb line. The intersection of these two lines will be the center of gravity of that part.

When forming long sheets in the press brake, the tail of the sheet needs to be lifted as the form-

ing progresses. You can use the crane to lift the sheet as it's formed, but I find it too slow and awkward (Figure 4-163). One solution is to use a simple lightweight block and tackle that can be suspended from the crane hook so the sheet can be hoisted to follow the forming. This sheet lifting is very important with sheets that extend far from the dies. Long sheets can be damaged if they are not lifted during forming. Simple C-clamps can be used to secure the sling to the sheet.

For heaver lifting jobs, I like set screw type c-clamps (Figure 4-164). These are the heavy duty burly brothers to normal c-clamps. They have a pointed set screw that is tightened with a wrench. The ease of attachment makes these clamps a versatile rigging option. They are a super secure easy to place lifting point (Figure 4-165).

Figure 4-161: Locating the center of gravity.

Figure 4-163: A crane is not always the best choice.

Figure 4-164:
Screw type
C-clamps.

Anti Gravity

In an earlier chapter I wrote that a small group of us worked late on Thursday nights on our own projects. This was typically a quieter time and place to think about your projects and try things you never got to do during the day.

One particular night my friend Doug and I were working late. It was past dark if I remember correctly. I was at one end of the shop and Doug was back working at his bench. The shop was pretty quiet as we were the only ones there. The rule was that two people had to be there for "safety" rea-

sons. The truth be told: Doug had the key and he came and went as he pleased. I just happened to be there that night.

I was in the middle of a problem or layout when I heard Doug calling my name. At one point he was the shop foreman, but at this point in time he was close to retirement and had given up that job. He still had the command voice however. I hollered back something like "okay, give me a second" or something along those lines. When he had called my name, it was in a calm way with no specific reference to time.

The second call was much more insistent and carried the command tone to it. Like a good puppy, I came running after I heard that one. As I rounded the corner, it took me a second to figure out what the heck was going on. I turned my head the way a dog does when it hears a funny sound it doesn't understand.

Now picture this: Your friend (the former shop foreman) is hanging upside down from the bridge crane. Got the picture? Me too!

Back in the 1980s, everybody was doing all these anti-gravity boot exercises where they put on special boots with hooks on the back and hung

Figure 4-165: These clamps are super secure.

Figure 4-166   30-ton main crane in the
Hunters Point Naval shipyard electronics shop.

upside down to do some kind of diabolically difficult routine. They even sold special tilting racks so you could hook the boots, then invert yourself in the privacy of your own home. Heaven forbid anybody seeing you upside down with your shirt over your head and your paunch lapping at your chin in the center of a health club!

True to his economic style, Doug had just used a rope sling with a couple of choker hitches around his ankles. He even had a neat little square of cardboard on the floor so he wouldn't get dirty rolling around on the welding shop surface.

There he was, hanging upside down, gently swaying back and forth a little on a Thursday night with nowhere else to go. In his most foreman-like tone, he barked out

"Get me down from here! C'mon, be quick about it!"

I noticed that he had lost his grip on the crane pendant and it had drifted away and out of his reach. For a split second, I thought about raising him higher while yelling, "The button's broken; I can't lower!" But, I quickly thought better of it because of the implications that the remainder of my life would be short. He had been unsuccessfully trying to swing himself back and forth and get enough momentum up to reach the stay pendant. This was where my services were required.

It seems Doug had strained his back earlier in the week. He had sneaked into the shop and inverted himself with the crane to stretch out his back, with positive results. This time, however, it seems he got a little sloppy and lost his grip on the control pendant. I reached for the pendant and slowly lowered him back on to his cardboard landing pad.

"How's the back doing?" I asked with half a chuckle. I was thinking to myself it was a good thing I was there or he really would have been in a bad mood Friday morning.

I hope you found a few tidbits in this section on rigging and lifting that might save you on some heavy lifting. Remember: go slow and be safe when handling loads heavier than you would care to drop on your foot. This is one skill you don't want to learn the hard way.

Manual Lathe

The manual lathe is the cornerstone of any machine shop. Almost all workers starting out in the machine shop finds themselves on the manual lathe. Lathes come in all sizes and shapes, as you can see in Figure 5-1. It has been called the King of machines for good reason. You can bet that if you need a forklift and a ladder to put the tool bit in the machine, some fun is bound to happen.

Anybody who has spent time on a modern lathe would immediately recognize all the design features of Henry Maudslay's revolutionary screw-cutting lathe, which he built around 1800. It is one of the oldest machine tools in which the look and features have not changed much since its invention. Another famous Englishman Joseph Whitworth added the compound rest feature which transformed Maudslay's design greatly. This was a major design breakthrough for what we now recognize as the modern engine lathe.

I started on the lathe in high school. At the time I was disappointed; I had wanted to be assigned to the welding section because I had been welding for a quite some time and was eager to demonstrate my skills.

The high school shop had four little Rockwell 9- or 10-inch manual lathes set off to one side of the shop. These lathes had the old rocker style tool posts and quick change threading gearboxes and taper attachments that took a little head work to figure

out. The funny part is that I somehow spent the entire semester on the lathes. I learned a tremendous amount and ended up enjoying the work immensely. Our first project was a stylish aluminum meat tenderizer, a definite must have for every modern kitchen or crime scene. We got to straight turn, taper turn, thread, and knurl on this one project. Like almost all high school machines, the corners of the compound rest were hammered to death from running them into the spinning chuck jaws. Later in life I found a great solution to this problem.

The shop teacher fabricated aluminum blocks that were glued or screwed to the compound and

Figure 5-1: Large Niles Engine lathe in the Mare Island Naval Shipyard Machine Shop. Check out the sign above the lathe.

served as sacrificial beating blocks for the lathe newbies. This improvement increases the resale value of any lathe. Among the first things used machinery shoppers look at are the condition of the ways and the corner of the compound. A clean crisp corner on the compound is usually an indication of a gentle life.

A few years later I got my first lathe, an old 1915 Prentice with a 9-inch swing. It had a flat leather belt that made a unique tick-tick-tick sound as the joint in the flat leather belt passed over the sheaves. It came to me with a huge stack of change gears for threading. If you want to learn to run a lathe, start on something old and loose. When you can get things done on an old worn machine, you will be a superstar astronaut on a tight modern machine. It took me a week to figure out how the clever little planetary back gear setup worked. The guy I was working for "offered" to let me store it for him while he was going through a divorce. I kept it for a couple of years and then ended up buying it from him for three hundred dollars. This was a huge sum because I was only making three dollars an hour part time.

Limp lead screw

A funny thing happened to me with that old lathe that took me years to get over. One day I was using the machine and the cross feed screw just gave up the fight. It had been getting pretty crunchy, so it wasn't totally unexpected. I took the slide apart and looked at the remains of the screw. At the ends it was fine, but the center section was obliterated. I decided to make a new screw and nut for it. I had access to another lathe at the time, so that decision was easy. I carefully measured the screw and nut. I decided to upgrade and make the screw out of stainless steel. I even remember the thread. It was a 7/16-20. After a few evenings of fussing around, I had my new lead screw and cool bronze lead screw nut. I was extremely proud of my downright cheapskatedness and felt a head-swelling self sufficiency about this time.

I put the whole thing back together, anxiously wanting to try out the new setup. I put it all back

together fine and my head began to swell noticeably. At least all my measurements were good. I went to spin the cross feed handle and got a big surprise when I realized the thread was supposed to be left hand. I had made the new screw a right-hand thread. Once I realized my mistake, the visual difference in the screw was obvious. Whoops! I really didn't feel like taking it apart again and making a whole new lead screw and nut, so undaunted, head still swollen with pride, I left it in and started using it.

Big mistake.

It took me a while to get used to the weird backward direction, which should have been my first clue I was headed for trouble. But after a while, it was second nature. The real problems started when I went to work in my first machine shop. I must have gouged a zillion parts before I unlearned the right-hand lead screw.

The moral of the story is: Don't learn anything too weird. You never know what you might have to unlearn.

I sold the lathe to a friend of mine for the same price I paid for it. I bet that same lead screw is still in that lathe. Poor fellow. . . .

On to the manual lathe section.

Small work can be done on large lathes but large work cannot be done on small lathes. Buy your lathe a little bigger than you think you might need.

Put target score marks on raw material for fast roughing (Figure 5-2). Touch the tool and use your scale to put reference marks so you can save

Figure 5-2: Target score marks.

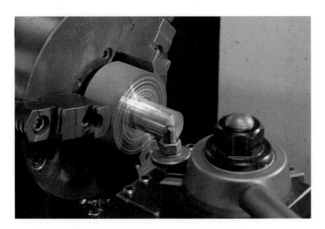

Figure 5-3: Rough cut near the lines.

measuring time. Rip and tear down close to the lines, then pull out your finer measuring tools (Figure 5-3).

Use quick change type tool-posts. Make sure you have plenty of tool blocks. I hate it when a lack of tooling interferes with my productivity. With these nifty, new, silver metallic sharpie markers, you can write your target numbers on the tool holder instead of a paper cheat sheet.

Always leave a boring bar set up in a tool block. Half-inch or five-eighths is a good starting point for general purpose work. Use inserted tooling if many different people use the machine.

Always have an inserted turning tool set up in a tool block for general purpose work if a few different people use the machine. CCMT or WNMG inserts are a good compromise (Figure 5-4). You can switch insert geometry easily for different

materials. WNMG (left insert of Figure 5-4) gives you six cutting edges per insert for good economy. We set ours up with the thought of one optimized for harder materials and one for softer materials.

Try to set up your dedicated turning tool block so that it misses the quill of the tailstock when the tool point is on center. This saves having to reposition the tool when using the tailstock to support your work.

Keep two parting blades set up: one neutral and the other with a couple of degrees left hand angle.

Keep a long, double-ended 45-degree chamfer bit set up all the time at each lathe. It is extremely handy for edge breaks and quick facing (Figure 5-5). The double end allows you to use it on both axes ID and OD.

Never modify someone else's hand ground tool left in a tool-block. You might as well ask to borrow their toothbrush. You just don't do it. If somebody leaves one in a tool block, take it out and leave it on the top of the lathe.

Always leave the lathe in better shape than when you found it. This has the added benefit of highlighting the shop slobs. They stand out in stark relief against a clean background, where they can be properly whipped and chastised.

Use a depth of cut (DOC) that is a little larger than the tool nose radius. With inserted tools, the chipbreakers do not function unless the depth of cut is larger than the nose radius.

To minimize chatter, use positive tool geometry with small nose radii and lead angle near ninety degrees. This is especially true for ID

Figure 5-4: CCMT and WNMG inserts.

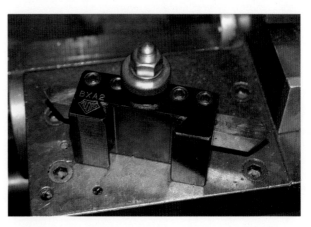

Figure 5-5: Double-ended, 45-degree chamfer bit.

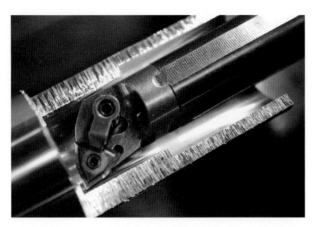

Figure 5-6: Step boring.

Figure 5-8: A boring bar shank.

boring. The tool tip should be on center or a few thousandths high.

Step bore deep bores in two or more steps (Figure 5-6). This leaves more room for chip evacuation and you can use a boring bar that fills the bore more completely (Figure 5-7). This is a very useful trick in the CNC lathe.

Carbide or heavy metal boring bar shanks are much more rigid for those deep holes. You can buy the heavy metal material and make your own boring custom bars. The long one in Figure 5-8 uses old broken 1/4-in carbide end mills for tool bits. Every shop has an endless supply of broken 1/4 tools that can be re-used in this boring bar. The longer bar in Figure 5-8 is a rigid Tungsten-heavy metal alloy called "No-Chat." It can be machined into any configuration you can imagine and lives up to its reduced chatter advertising.

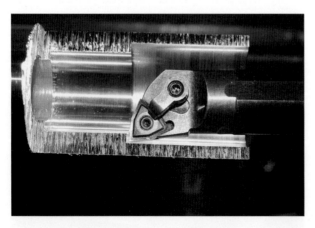

Figure 5-7: Using a boring bar.

When you have a chatter problem, try increasing the feed rate before you try slowing everything down. Another trick is to move the boring bar in the holder a fraction of an inch in either direction. Sometimes this small change in the resonant frequency can reduce or even eliminate chatter.

Carbide inserts have a calculated design life of twenty minutes of cutting time. This is great if you sell inserts and not so great if you have to write the checks for them.

Always try to increase cutting speeds and feed rates. If you never push the envelope, how do you know where the limits are?

A 20% increase in feed rates returns a greater reduction in part cost than a 50% increase in tool life.

Test and try new tools once in a while. Lots of smart people are working on some really good tools. Besides, it's fun to test the salesman's tools at full throttle.

Be aware of the reduction in clamping force when spinning a chuck at high speed. You can lose 50% of your clamping force at high RPM.

Orient rough cut blanks from oxy-fuel, waterjet, and plasma cutting with the large end of the taper in the cut to the back of the chuck (Figure 5-9). All of these processes produce varying amounts of taper, depending on the material thickness (Figure 5-10).

Interrupted cutting. About the only thing good about interrupted cutting is that the chips break no matter what.

Figure 5-9: Orient rough cut blanks.

When roughing, try to get completely under the bark or outer skin of a bar on the first pass. This will be repaid in tool life.

Most lathes have .001-diameter divisions on the cross slide dial. A decent operator can control diameters to .0003 by interpolating between the divisions—if the lathe is good shape.

With that said, your lathe should repeat to half the smallest division on the cross slide dial.

You can set your compound rest at a small angle to dial in increments smaller than the cross feed dial dimensions. A 5.73 degree angle off the Z axis will take off .0002 on the diameter on the X axis for each .001 dialed on the compound. When I have used this trick, I set it at an even 5 degrees to keep me on the MMC side. This is an old trick that is seldom used in the job shop, but I still like the idea.

Figure 5-10: The amount of taper produced varies.

Figure 5-11: An internal expanding collet.

Figure 5-12: Internal expanding collets hold bushings and sleeves.

Try to grind your special lathe tools so the tool post can stay square to the machine axes. This saves setup time, but is not always possible.

Hardinge makes excellent internal expanding collets (Figures 5-11 and 5-12). These are great for holding bushings and sleeves on the inside diameter. One inner mandrel is used for a large range of expanding collet diameters. They are still soft enough that you can turn the OD a little for one of those oddball emergency jobs that always seem to come up.

Try this trick for holding on the ID of plastic parts or bushings. You can expand them a few tenths by dropping them into hot water (Figure 5-13), then slipping them on to a gage or dowel pin (Figure 5-14). You can then turn a precision OD by holding the pin in a collet or chuck. To remove the part,

Figure 5-13: Drop plastic parts or bushings in hot water.

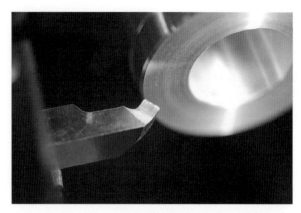

Figure 5-15: Grind multipurpose tool bits.

Figure 5-14: Slip parts or bushings on a gage or dowel pin.

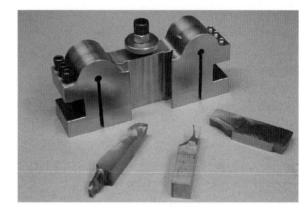

Figure 5-16: A spring-type tool holder.

just dip the pin and bushing back in the hot water. By exploiting the fact that the plastic expands much more than the steel and faster, you can slip it off the pin.

Grind a couple of multipurpose tool bits (Figure 5-15). This can save time on manual tool changes. The one in Figure 5-15 turns ODs, faces, chamfers OD, and ID—all in one tool.

Make yourself an old school spring-type tool-holder (Figure 5-16) if you need to do large radii or use form tools (Figure 5-17) with a broad cutting edge, or any tool that you are having chatter problems with. The spring-type tool-holder with its pivot above the centerline backs the cutting edge off when the tool bites and starts to chatter. Typically you can increase your cutting speed by double, using a setup like this. Contrary to the normally accepted thinking, sometimes more rigidity is not the answer.

In the old days when planers were used to surface plates and other large flat surfaces, the final finishing tool was a very wide flat tool held in a gooseneck or spring-type tool holder similar to the one in Figure 5-18. Form tools are an excellent and fast way to duplicate complicated geometry in the manual lathe.

Figure 5-17: Use form tools.

Figure 5-18: A very wide flat tool.

Figure 5-21: Another look at the same form tool.

Figure 5-19 shows a curved form tool that on a good day has chatter written all over it. In this example, this tool was plunged into 4140 steel (Figures 5-20 and 5-21).

For those special jobs, you can have your local wire EDM shop cut you special profiles and difficult-to-grind geometries. Figure 5-22 shows a couple of examples we have done over the years. Some of these would have been a nightmare to hand grind accurately. Try finding commercially-made tooling for some of the crazy stuff designers come up with.

Pull back on the tool post with a few pounds of force when backing your tools in the Z axis (Figure 5-23). This prevents leaving tool tracks in your turned surface. This works with boring bars on the ID also, but you have to push instead.

Figure 5-19: A curved form tool with chatter.

Figure 5-22: Special profiles.

Figure 5-20: This form tool was plunged into 4140 steel.

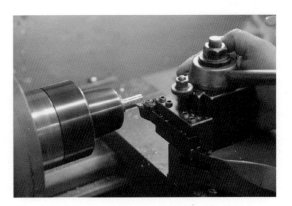

Figure 5-23: Backing tools in the Z axis.

Figure 5-24: Making double ended parts.

Save on holding stock by turning double ended parts, then cutting them off at the centerline (Figures 5-24 and 5.25).

Make yourself a couple a quick aluminum face plates that fit in the three-jaw chuck (Figure 5-26). These can be re-surfaced dead flat dozens of times and are quick to set up. These two are welded together, but they can just as easily be bolted. If you bolt them, be sure to sink the heads well below the surface so you don't face the screw heads.

A nifty hand-tapping guide that fits in the tailstock chuck works well for hand or slow speed, hand-tapping small thread sizes (Figure 5-27).

Three-jaw backing plates are great for backing up a part for heavy drilling (Figure 5-28). If you make several thicknesses, they can be used to position thin disc-shaped parts quickly (Figure 5-29). Add three-jacking screws for adjusting the plate in

Figure 5-26: Aluminum face plate for a three-jaw chuck.

relation to the chuck face to hold thin disc-shaped parts right where you want them in the jaws against a nice parallel surface.

You can use light cuts for a disc-shaped part by pressure plating against a face-plate (Figure 5-30). You can even use the top of the chuck jaws if you have a little help. Three pieces of double stick tape

Figure 5-25: Cutting double ended parts at the centerline.

Figure 5-27: A hand-tapping guide.

Figure 5-28: A three-jaw backing plate.

Figure 5-30: Pressure plating against a face plate.

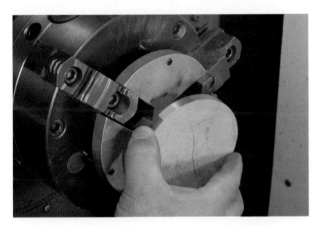

Figure 5-29: Positioning thin disc-shaped parts.

make all the difference (Figure 5-31). Open the jaws to a radius just below what you will be turning to for maximum holding. Use a smaller disc with a center hole in it to push against with the tailstock center (Figure 5-30).

This method works great when you can't have a center drill or center mark in the work piece. Double stick tape always works better when the mating surfaces are cleaned with alcohol and the blank is squeezed into the tape.

The Russian Spy

My old toolmaker buddy worked in the naval shipyard near Bremerton, Washington, during WWII. He told me quite a few great stories about the day-to-day life in a machine shop during the big war. This was one of those big government

shops with hundreds of guys on multiple shifts, building everything from torpedoes to potato peelers.

When my friend Charlie first started in the shipyard, he was assigned to the lathe department. They had quite a few lathes with a huge range of sizes. He started out on a lathe with a very long bed, turning long shafts and screws, I think he told me that he sometimes had to mount up to four or five steady rests at a time for some of these superlong shafts he had to make.

After he had been working there a while, he got a reputation as a pretty good lathe hand. One day, eight or ten new lathes were delivered to the shop and set up. American Pacemakers, he told me, "Good American machines," as he recalled the details to me. Being a relatively new guy, he was surprised when he was assigned to one of the new

Figure 5-31: Getting help from double stick tape!

machines. *After a little grumbling from the older guys, things settled down.*

Nearby on another lathe, a machinist set up his machine to do some drilling on a part. Apparently it was a pretty large drill bit on the order of two or three inches in diameter. The guy set the machine up and started drilling. I guess the hole was pretty deep so he was at it for some time.

At one point there was a big commotion at this guy's machine and everybody came over to see what had happened. Like sharks to a scuba diver with a beef jerky wetsuit, they descended on this poor fellow. When my friend Charlie got over there, everybody was still laughing at this poor red-faced guy for what he had done.

Apparently he was daydreaming while he had to drill the hole and didn't notice that the chips coming out of the hole had changed color. He had drilled completely through his part and kept going all the way through the four-jaw chuck. Everybody was laughing because the chuck was hanging on the big old drill bit and had been drilled off the end of the spindle.

Now I have to admit I would have been cracking up myself. On one side, I feel sorry for the guy but, damn, that's some funny stuff. Everybody was laughing at this guy's pain and humiliation when Charlie piped up and said, "Hey, I think this guy is a communist spy." I guess people quieted down for a second to address this serious accusation just long enough for Charlie to say,

"He's a Russian spy, I tell you. His name is Borhis Chuckoff."

The place erupted in a fresh round of laughter and a nickname was born. I think this guy was on the permanent harassment list from that point on. I don't think anybody even remembered the guy's real name after that. "Hey Borhis, how's little Borhis? Hey Borhis, what did you bring for lunch today? Chuck soup."

For repetitive eccentric turning, clamp a small three-jaw chuck or collet block in your four-jaw chuck (Figure 5-32). Nobody likes to dial in every single part in a four jaw.

Figure 5-32: Repetitive eccentric turning.

Turn the ID and OD taper with the same compound setting when trying to match tapers exactly (Figure 5-33). Run the lathe in reverse and cut on the backside for one of the tapers (Figure 5-34).

When turning tapers with a specific end diameter, make yourself a couple of quick gages. Taper diameters are difficult to measure accurately in the

Figure 5-33: Turning the tapers.

Figure 5-34: Running the lathe in reverse.

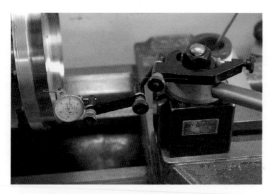

Figure 5-35: Flexible indicator holder.

Figure 5-38: Square-to-round parts.

machine. A little trig and a simple ring gage will make your life easier.

Use your flexible indicator holder on the top of the lathe tool post (Figure 5-35). There is just enough of a lip for the clamp to grab.

A test indicator on a compact magnetic base is quick to set up, which makes you use it more often (Figure 5-36).

Hold large square parts outside the range of collets in a round sleeve (Figure 5-37). They are

easy to make and are faster than a four-jaw chuck if you have a few square-to-round parts to do (Figure 5-38). A real organized person might make a few of these up for common square sizes.

Add a short counter-bore .001-.002 larger than the size of the part to be pressed in Figure 5-39. This makes assembly easier because the part is started dead straight. An alternate is to turn the male part down for a perfectly straight start (Figure 5-40).

Figure 5-36: A test indicator.

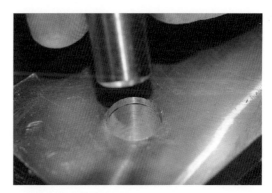

Figure 5-39: Adding a short counter-bore.

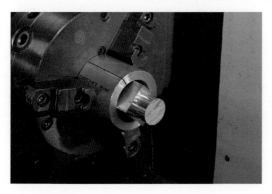

Figure 5-37: Holding large square parts.

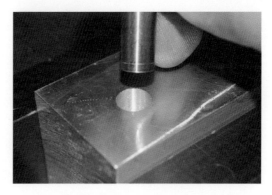

Figure 5-40: Turning the male part down.

Figure 5-41: Pressing in bearings and bushings.

Figure 5-43: Aligning the parting tool against the chuck.

You can press bearings and bushings in with the lathe tailstock (Figure 5-41). They will go in straighter when pressed this way. Don't worry! The lathe and the mill can handle the thrust (Figure 5-42). What do you think happens when you're leaning on a big drill.

Align your parting tool against the flat face of the chuck (Figure 5-43). You can see tiny variations easily and it's super quick.

The quick change tool-post can be aligned in the same manner.

Try power feeding your parting tool. Make sure it's sharp and square with the lathe axis. Keep it wet and start out with .002/Rev feed. The constant steady feed pressure works wonders. Most people have had trouble with parting at one time or another, so they tend to run too low a cutting speed and feed lightly. It you doubt me watch a CNC lathe

part stuff off; it's too dumb to be scared of parting, so it just does it.

For large diameter parting, you can part in several steps by moving the blade out in increments in the blade holder. Don't forget to adjust your center height when you move the blade in or out.

Turn true bands on tubes and shafts to get them to run true in a steady rest (Figures 5-44 and 5-45). If you don't turn a concentric band, the steady rest will just follow the out of round of the tube or shaft, producing a new surface to match. If you cannot cut a true band on the outside because of the final part geometry, you can clamp a sleeve on the outside and turn that true for the steady rest to run on.

When turning small part diameters, make sure your tool is dead on center and razor sharp.

Put a disc or a spider in the end of a tube that's too big for the steady rest (Figure 5-46). A bronze

Figure 5-42: The lathe and mill can handle the thrust.

Figure 5-44: Turning true bands on tubes and shafts.

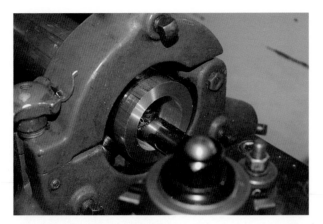

Figure 5-45: Getting bands to run true in a steady rest.

Figure 5-46: Putting a disc or a spider in the end of a tube.

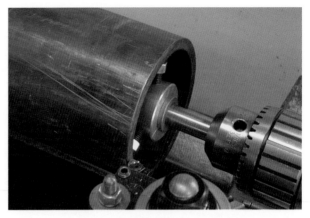

Figure 5-47: Providing the pivot.

start the spindle when using the tension block. The wire centerline should be at the lower tangent point of the winding mandrel. Did I mention to make sure you run the lathe slowly? Turn the spindle on before you try this, to confirm the speed setting.

Figure 5-48: The quick vise grip method.

bearing and a shoulder bolt or a dowel pin in the drill chuck provide the pivot (Figure 5-47). This allows you to face the entire end of the tube square supported on center. You can part off big rings using this trick also. Just be sure you're parting off on the right hand side of the spider.

When winding springs or rings in the lathe, you can use the quick vise grip method (Figure 5-48) or, if you want more consistent results, make a tension block that fits in the tool-post (Figure 5-49). The factors that affect finish ring or spring diameters are mandrel diameter, wire condition, and initial tension. The vise grips are special homemade with the teeth removed and smooth copper jaws silver soldered in place (Figure 5-48). You don't need that much clamping force to hold the wire in tension as you wind it. Run the lathe at its slowest speed. Then be sure to engage the feed before you

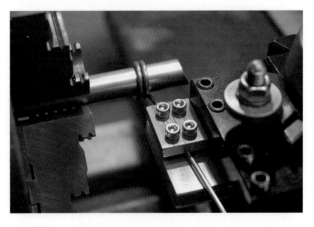

Figure 5-49: A tension block.

Figure 5-50: Squeeze the coil together.

Squeeze the coil together and cut all at once for the best results (Figure 5-50). Twist the legs gently into alignment and weld for professional results (Figure 5-51).

I keep several test indicators set up differently (Figure 5-52). You can jump from the height gage to the drill chuck to the spindle nose quickly without screwing around with all the little pesky clamp fittings.

Figure 5-51: Twist the legs gently into alignment.

Figure 5-52: Test indicators.

Figure 5-53: A spin handle.

Make yourself a spin handle for the compound rest if you hate turning short tapers as much as I do (Figure 5-53). You will get better finishes on your compound tapers and save your wrists on the heavy feeding.

You can use the chuck jaws on the lathe for a quick, accurate 120-degree layout (Figure 5-54). A block placed between the jaw and the way makes an accurate stop while you scratch a line with a pointed tool (Figure 5-55).

If you need multiple start threads, pick three-start if you have a choice. It's easy to index the shaft three times using the jaws of a three-jaw chuck as a reference surface.

Simple drive dogs can be quickly made from stock shaft collars (Figures 5-56 and 5-57). Turning between centers is still the most accurate way to establish concentric diameters end to end (Fig 5-58).

Figure 5-54: Using the chuck jaws on the lathe.

Figure 5-55: A well-placed block makes an accurate stop.

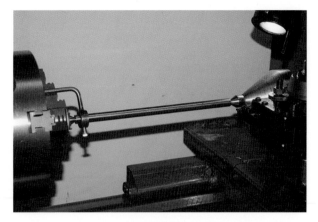

Figure 5-58: Turning between centers.

Turn a soft center held in the three-jaw at 60 degrees included angle (Figure 5-59). It will be dead accurate on center if you don't move it after you cut it.

Set up a simple stop pin in a tool-block to save all your tool offsets on the DRO. Pull your stock out of the headstock against this pin to accurately set your starting position.

Put a small rod in the drill chuck to catch multiple small parts when parting off (Figure 5-60). It beats scratching around in the chip pan for your parts.

If you have a job where it's not an option to lose the part, do yourself a favor and clean the machine and chip pan before you start. You can also lay a sheet of brown paper in the pan so you can easily see what falls in.

Figure 5-56: Simple drive dogs.

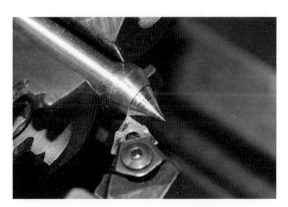

Figure 5-59: Turning a soft center.

Figure 5-57: Using stock shaft collars.

Figure 5-60: Catching small parts.

Figure 5-61: Measuring the spindle centerline.

Figure 5-63: Using a tiny spot as the target.

Keep a square bar near each lathe long enough to span across the apron and accurately measure the spindle centerline distance off the bar (Figure 5-61). This distance is permanently engraved on the bar. It makes it easy to set every kind of tool height no matter where the cutting edge is. On most lathes, the apron surface is the same height on both sides of the tool-post.

As an alternative, you can use the Sharpie centering method (Figure 5-62). Leave a tiny little spot on the face of a spinning part and set your tool to this target spot (Figure 5-63). Presto! Instant center height!

Make yourself three copper sheetmetal jaw covers for delicate work (Figure 5-64). Be sure to pre-clamp the new jaws on a test piece to set the final curvature (Figure 5-65). Safety wire or nylon ties keeps them from dropping out every time you open

the chuck. Or you can bend ears over like these. Material is .03 copper.

You can plunge turn small, long-aspect ratio diameters by taking the entire cut in one shot (Figure 5-66). Be sure the starting stock diameter is large enough to handle the extra cutting pressure without deflection.

You can also step turn this kind of part. Your chuck or collet must run very true for best results.

Figure 5-64: Copper sheet metal jaw covers.

Figure 5-65: Preclamping the jaws.

Figure 5-62: The Sharpie centering method.

Figure 5-66: Taking the entire cut in one shot.

Figure 5-68: Clamping the rod.

The shaft is pulled out a little at a time as the diameter is reduced in steps along the z axis. This works best if the stock material is centerless ground for low runout.

Make yourself a retractable tailstock stock stop (Figure 5-67). When you have a bunch of parts to part off to a specific length, this works great.

You can use stock stop in headstock for repetitive chuck work. Clamp the rod with the screw in the middle of the bushing (Figure 5-68). The set screws in the counter-bore lock the bushing into the spindle bore (Figure 5-69). It uses a cut-down allen wrench to get at the screws. I don't really like this approach, but it does work and it is used so infrequently in our shop that nobody wants to spend the time to make a nifty one. Ideas, anyone?

When you need to cut a dovetail o-ring groove, set your compound at the half angle of the dovetail

Figure 5-69: Locking the bushing into the spindle bore.

and cut along the angle with the compound (Left, Figure 5-70). This allows you to use a larger, thicker tool than if you plunge in straight perpendicular moves for a job that on a good day can be difficult. The usual angle off the lathe centerline is 24 degrees.

Long stock guide. Whenever you have long stock sticking out of the headstock (Figure 5-71), you need to be **extremely** careful. It's best to cut

Figure 5-67: A retractable tailstock stock stop.

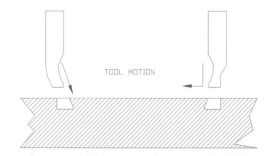

TOOL MOTION

Figure 5-70: Setting your compound at the half angle.

Figure 5-71: A long stock guide.

Figure 5-73: Soft jaws.

the stock so it fits within the spindle. But, if you must have it stick out, use a guide and run the spindle SLOWLY. Much shop embarrassment, injury, and dental work have been caused by this specific operation. I have personally seen 2.5-diameter plastic whipping around at 3000 rpm and actually break off. Talk about run for cover. Check your spindle speed setting **before** you put the stock in. Did I mention this is a squirrelly operation?

Wind up a piece of thin, soft plastic sheet to keep the bars from getting scratched on the inside of the stock support tube (Figure 5-72). The plastic springs open when it's inserted in the tube, which helps retain it while the work piece turns inside. And for goodness sake, run the machine slowly!! Did I mention this is a very dicey operation and should only be used if you are in a bind?

Here are some really easy-to-make soft jaws that can be rotated and have four usable edges

(Figure 5-73). Use these for light-to-medium work, like boring sprockets accurately on the pitch circle (Figure 5-74).

Make yourself a couple of soft jaw spiders from large hex nuts (Figures 5-75 and 5-76). They will cover a million diameters and save hunting down the right disc in the scrap bin. You can easily set them with calipers (Figure 5-77).

The Donut Deflector

A few years ago, I was in the field at one of our customer's facilities installing some equipment we had built. I won't name any names, but it was one of the big government research labs. I noticed when we came in that off to one side was a small machine and mechanical workshop to support the activities in the equipment bay where we were

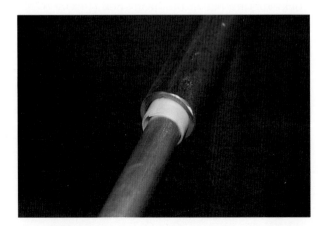

Figure 5-72: A plastic sheet protects the bar.

Figure 5-74: Boring sprockets.

Figure 5-75: Soft jaw spiders.

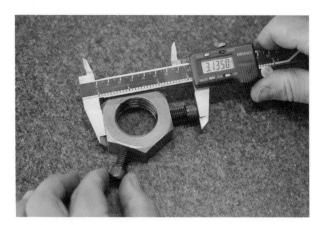

Figure 5-77: Setting soft jaw spiders with calipers.

working. At one point we stopped to take a break and talk with the technician we were working with.

Now if you have not figured it out yet, I am a huge gear-head through and through. I love looking at machinery and, if there is an opportunity for getting a shop tour, I always ask. So I asked our escort if he would give us a quick tour of the shop. He readily agreed and showed us the way.

It was a nice little model shop. We were not allowed to take pictures or even have a camera with us so I have no pictures of this place to share. This shop was very well equipped to help the engineers and scientists working in the bay do pretty much whatever they needed to have done. They had a little sheetmetal brake and a Rotex punch for making little brackets and guards as well as a complete machine shop.

About half way through the tour we got to the manual lathe they had in the machine shop.

Figure 5-76: Using large hex nuts.

Nothing special, just a good plain Jane, American-made engine lathe. I hardly looked at it except for one major difference.

Just to set the stage a little: this is a government laboratory where they handle and work with nuclear materials routinely. You're almost afraid to ask questions because you may be prying into the forbidden world of national security and black ops secrets. Much of the work is strictly classified and top secret. The fact that we had an escort assigned to us while we worked in a mundane, unclassified area gives you some idea.

On the rear of the headstock of this lathe was some kind very special stainless fitting. I have seen many lathes and the purpose of this peculiar fitting was definitely not obvious. It was a very serious piece of hardware. More like what you would see on a vacuum chamber or an expensive piece of laboratory equipment. It was completely out of place on this plain non-descript lathe.

Immediately my mind was trying to decipher the secret of this fitting and why it would be on a simple machine like a manual lathe. The fitting consisted of a thick stainless flange attached directly to the machine. To this was welded a short length of tube which was then welded to a ninety degree tube ell which terminated in a fancy, high-vacuum Conflat flange. The whole thing was electropolished to gleaming perfection. Some of the uses I came up with in my mind involved extracting chips and fumes from noxious radioactive materials, or perhaps some kind of innovative, vacuum-holding setup. In this lab, the possible uses were mind bog-gling. It could almost be anything. Finally I asked our escort if he knew what this beautiful piece of gleaming stainless steel was used for.

He said, "Oh that thing. It's a donut deflector."

Now I was really curious. What the heck is a donut? Some kind of code name for a super-secret part that dwells deep inside the core of a nuclear weapon? An experimental magnet part for a phased plasma fusion reactor. What the heck was a donut and why did it need to be deflected?

So I asked the guy, "What's a donut and why does it need to be deflected? How does it work, if you can tell me?" I didn't want to probe or put our escort in the position of being downrange of a fir-ing squad for betraying any classified information by answering my barrage of questions.

He laughed and said, "Well, the donuts don't really get deflected. You see Billy Bob brings a dozen of the best glazed donuts you ever tasted every Wednesday. He always puts them on the desk here." He pointed to a spot on the desk directly in line with the lathe spindle. "One time some moron was running the lathe and blew a bunch of chips down the spindle bore and they shot out and got all over a steaming fresh batch of donuts. Wrecked the whole batch," he said with a tear forming at the corner of his eye. "We built this deflector on the lathe here to keep any chips from landing on the desk and the donuts."

All I could manage was a weak, "Oh." You know that sound an air mattress makes when you pull the plug and roll it up. "Pfffhhhh," that's how I felt. We left the shop and finished our work. All I could think about was the thousand bucks in stainless fit-tings and an unknown number of man-hours for the sake of a dozen donuts. I kick myself now for not asking the guy where these magnificent fried pastries came from. . . .

Step turn large radii in the manual lathe using x and z coordinates. You can cut large radii in the lathe using a method I call step turning. If you don't have a CNC lathe, this might be the only option for large curves. This is one of those things that is much easier with the assistance of a computer to do

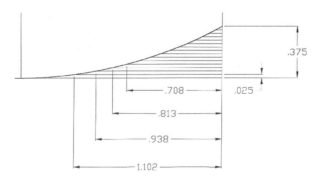

Figure 5-78: Cutting a large radius.

the graphic layout and all the math. You can do the math manually but it takes a little longer.

The first step is to lay out the desired curve accurately at full scale or, if you're doing the cal-culations manually, maybe ten times scale for more accurate final results.

In this example, we are trying to cut the large radius in Figure 5-78. We can do this by taking a series of cuts of a certain depth that all have their z endpoints on the desired curve. In this case, we are taking a radial depth of cut of .025 per pass. Each successive pass has a different z end point that ends exactly on the curve. Depending on the accuracy needed, you can take as many passes as you like. Don't make the radius diameter mistake. In the above example, we are only showing half the curve, but removing .050 on the diameter per pass.

Figures 5-79 and 5-80 show a lower anvil for a large English wheel machine. You can see both

Figure 5-79: Lower anvil of a large English wheel machine.

Figure 5-80: The sides are ready for finishing.

Figure 5-82: Removing excess material is the next step.

sides have been step turned and are ready for the finishing. If you're doing the layout manually without the benefit of a computer, measure your points on the curve from a baseline with calipers. It helps to draw your curve 10x scale for better accuracy.

In Figures 5-81 and 5-82, you can see where the curve is partially blended in. We already know the corners of the steps lie directly on the actual curve, so all we need to do is remove the excess material between each pass. If you take a large number of passes, the material left to remove can be as small as you like. Also, the trailing edge of the tool can be shaped to help remove the excess between the step lines.

Figure 5-83: Using a rotary type radius tool.

You can use a rotary type radius tool in the lathe instead of grinding one from scratch (Figure 5-83).

Threading in the Manual Lathe

Screws and screw threads literally hold the world together. There are nearly as many types and forms of threads as there are products that use threaded fasteners and connections. Equally there is much confusion and misuse of threads in general for the non gear-head genre.

From the machinist point of view, cutting threads is kind of a satisfying experience. When you're done, hopefully you have two parts that mate together with a precision and smoothness you

Figure 5-81: A partially blended curve.

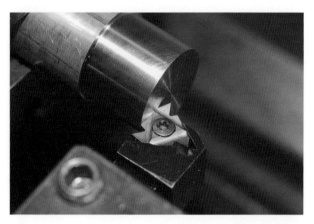

Figure 5-84: Aligning your tool.

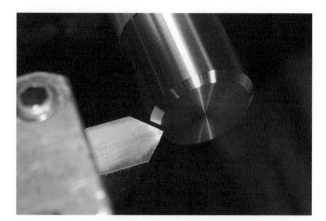

Figure 5-85: Starting your threads.

just don't get with run-of-the-mill hardware grade fasteners. I have always enjoyed cutting threads in the manual lathe and have learned a few tricks over the years that are of interest.

Align your tool against a freshly-faced end or against the side of the chuck (Figure 5-84). The little arrow-shaped alignment tools you see are a pain and are only good for gaging hand ground tool bits.

If you do a lot of threading in the manual lathe, invest in an inserted tool-holder. The inserts are ground with near perfect geometry and are easily changed. One insert cuts dozens of thread pitches.

When I learned how to thread in the lathe, I learned using the compound in-feed method. Contrary to some popular beliefs, you do not have to have the compound set at half the thread angle. By using what's called modified flank in-feed, and changing this angle, you can help alleviate threading problems with difficult materials.

Another advantage to threading with the compound is you don't have to keep track of your dial position. The cross-feed dial is always zeroed after each pass, so you have less to remember. For example, was that last pass at .030 or .050? The main disadvantage is your Z position changes as you feed in. This is usually not a problem on OD threads, but it can be a problem on internal threads that end against a shoulder.

How do you start your threads? Figures 5-85 and 5-86 show a couple of examples. I like the

chamfer with the threading tool (Figure 5-86). It saves a tool change. Be sure to chamfer a little deeper than the minor diameter of the thread.

How do you end your threads when the designer has not specified a specific relief? Here are a couple of ideas to try out. When I want to do something with the groove that gets cut at the end of the thread, I usually just use the threading tool and traverse a small relief at the end (Figure 5-87). It saves a tool change and looks okay. If I want a little nicer look, I sometimes switch to a radius tool (Figure 5-88). Just be sure you are a little smaller with the relief than the thread minor diameter so the mating part will thread all the way to the shoulder.

Use a large DOC on your first pass during threading. The point is small; in the first couple of passes, the area of the tool tip engagement is

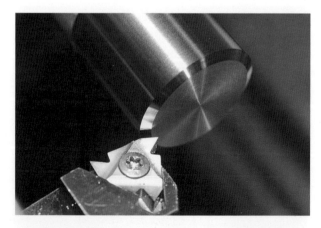

Figure 5-86: Chamfer with the threading tool.

Figure 5-87: Traversing a small relief at the end.

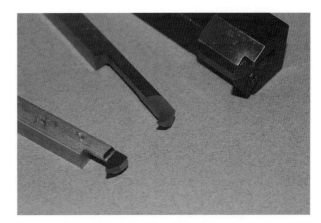

Figure 5-89: Left-handed tools.

also small. Taper off your depth of cut as you get deeper.

On your last pass, feed straight in with the cross feed a light .001 spring cut. This cuts on both flanks of the tool and literally cleans the thread of any chatter or tool marks.

I can never remember which line on the threading dial to use with which thread pitch. If you're lucky, it will be marked. When in doubt, just use the same number or line each time. Always use the same number when cutting multiple start threads.

Do your internal threading from the inside out with left-handed tools (Figures 5-89 and 5-90). You will get less chatter and you can see what's going on down the bore. You will need left-hand threading tools and run the lathe in reverse. The way it was explained to me was, "It's easy to pull a rope; it's really hard to push one."

Figure 5-90: Internal threading from the inside out.

When you have a choice, fine threads are easier to cut and need fewer passes. The shallower depth on difficult materials might save your bacon.

For quick and easy day-to-day threading gages, I keep a complete set of nuts in my toolbox for fitting threads (Figures 5-91 and 5-92). One ring hold coarse threads and the other ring holds fine threads. When you thread, be sure to run the nut the full length of the threads. When left to their own devices, machinists tend to cut threads tighter than necessary.

Mating materials in threaded connections are important. If you must use the same material for male and female threads, do yourself a favor and put a few molecules of thread lubricant or anti-seize on them before you crank them together.

Figure 5-88: Switching to a radius tool.

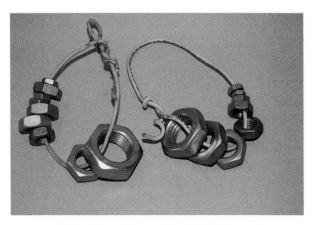

Figure 5-91: Keep a complete set of nuts.

Figure 5-93: A thread file.

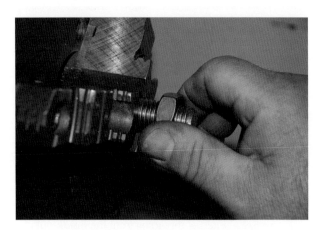

Figure 5-92: Run the nut the full length.

If you do happen to get your male and female threads wedged together in an intimate embrace, a simple trick to get them separated is to quickly warm the female portion 100 degrees or so, using a propane torch. A quick shot of penetrating lube before you twist and you might just save your work.

When measuring threads, a dedicated thread micrometer is very handy and quick at the machine. But for the highest accuracy, use the three-wire thread measuring method. The reason the three-wire method is more accurate is because the wires present a true parallel surface for measuring. If it's good enough for the gage makers, it's good enough for me.

A piece of modeling clay or window glazing putty can help you hold those pesky thread meas-

uring wires. Better yet, buy a set of the plastic holders that fit the micrometer spindle.

Thread files actually work (Figure 5-93). They are great for straightening the annoying half thread fade at the beginning and end of an external thread.

Multiple Start Threads

Here is an example of cutting multiple start threads, sometimes called multiple groove threads, in the manual lathe. These threads are used for getting a high lead per revolution with a shallow thread depth.

Suppose you have a .25-inch per revolution lead, but it is to be cut on a small diameter cylinder or thin-walled tube. The normal double depth for a .25-lead, 60-degree thread is .324. If you wanted to cut this on a shaft of .375 diameter, you would pretty much be out of luck. Enter the multiple start thread.

As the name implies, there is more than the normal single start. These threads can be identified by looking at the end of the thread and counting the number of entry starts you see. There is no practical limit to the number of starts you could do. The limitations typically are with the machinery used to produce them. Most engine lathes will not thread coarser than 2 TPI. In the CNC lathe, this is not the case, but that's another chapter.

Figure 5-94: Prepare the threading blank.

Figure 5-95: Prepare the levers, dials, and compound.

Let's work through the example I have done. The process is the same as cutting normal 60-degree threads we find all over the place with a few exceptions.

For multiple start threads, we must index or adjust our starting position for each separate start. This can be done in several ways. The first way is to index the part radially precisely the number of starts you wish to have. So if you have a three-start thread, you would index each start 120 degrees. Or if you wanted a four start, you would index your part ninety degrees. It's important to note that the axial or Z position cannot change at all when you use the part index method. This limits you to threading the part between centers to maintain the same z position.

The second method is easier, but only if your lathe has a compound rest attachment which will swing ninety degrees to be aligned with the bed of the lathe.

The first step is to prepare your threading blank. In the example, the diameter is arbitrary (Figure 5-94). We will cut a four-start thread with a lead of .25 per revolution. To determine the actual thread dimensions to cut the thread, we will divide the .25 lead by the number of starts. So we have (.25/4) = .0625. This corresponds to a thread with a lead of one-sixteenth per revolution or 16 TPI. This is the thread depth we will cut for each of the four starts.

Make sure your threading levers are set for the desired lead which, in this example, is .25 per revolution or 4 TPI. Remember we are cutting a 4-TPI thread, but with four individual 16-TPI depth grooves. I always start by zeroing all my dials cross feed and compound. Make sure the compound is set at 90 degrees and the dial is zeroed (Figure 5-95).

Out of old habit, I always take a .001 scratch pass to confirm my gearbox settings (Figure 5-96 and 5-97). You would be surprised how many times this gets fouled up. For high lead threads, you will want to run the spindle pretty slowly, especially if you are heading toward a shoulder like this example.

I always use the same number on the thread dial for multiple start threads (Figure 5-98). It is

Figure 5-96: Make the first scratch.

Figure 5-97: Confirm the settings.

Figure 5-99: The first groove at final depth.

most likely okay to use other lines or numbers, but who wants to screw up their work? Typically threading is one of the last operations, so you have invested some time to get this far. Why risk a failure? In Figure 5-99, I have cut the first groove to final depth.

Now for the second groove. The compound rest must be moved a distance along the z axis that is equal to the thread lead divided by the number of starts. In our example, $(.25/4) = .0625$. So we advance the compound .0625. The direction that the compound is moved does not matter as long as you don't change once you start in a particular direction.

I made a confirming scratch pass for the second groove to be sure I moved the compound the correct amount (Figure 5-100). In Figure 5-101, the second groove is cut to full depth After completing each groove, the compound is advanced the distance of (lead divided by number of starts) along the z axis.

After the third groove, it's starting to look like something (Figure 5-102). And after the fourth groove, it looks like a funny 16 TPI (Figure 5-103).

Figure 5-98: The thread dial.

Figure 5-100: The second groove at scratch pass.

Figure 5-101: The second groove at final depth.

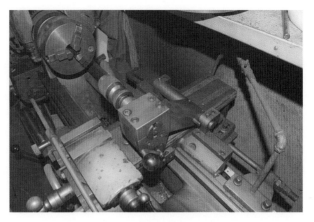

Figure 5-104: A hinged, flip-down drill chuck.

In fact, if you put a standard thread gage in the grooves, it should fit the 16 leaf. The only thing that looks different is the lead angle, which looks much steeper than a normal 16 TPI.

The process is fundamentally the same for an internal thread. If you find yourself having to do an

internal multiple start thread, be sure to try the inside out ID threading that was discussed earlier. When you make the mating female thread for this example, it will have a lead of .25 per revolution, but a thread depth of a 16 TPI. Pretty cool!

Figures 5-104 and 5-105 show a pretty interesting device I saw recently on an old lathe. It's a hinged, flip-down drill chuck attached to the apron. It has a clever latch to lock it into place on the machine centerline. This allows you to use the power feed for big holes and the hand wheel for rapid pecking on deep holes. It also allows you to have more than one centerline tool set up at once like a drill and a tap.

Here is a way to quickly couple your indicator to the tailstock quill (Figure 5-106). Sometimes it's important to get a precision depth with the tailstock. This method saves making a special bracket

Figure 5-102: The third groove.

Figure 5-103: The fourth groove.

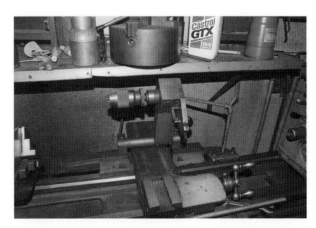

Figure 5-105: The chuck is attached to the apron.

Figure 5-106: An indicator coupled to the tailstock quill.

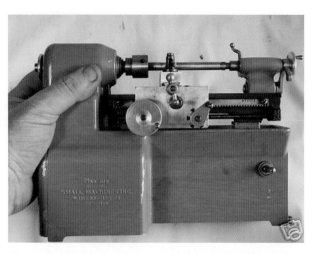

Figure 5-107: A tiny Manson pocket backpacking lathe.
Just kidding; it's not for backpacking!

that most likely only fits one lathe. You can count turns if you like, but I find this method less prone to daydreaming.

Manual Milling Machine

If the manual lathe is the king of machines, then the manual mill must be the queen. A shop is just not a shop without at least one manual mill. Years ago, before the current vertical mill configuration became readily available, the shaper was the shop heavyweight for prismatic type work. Ever since the introduction and evolution of the modern vertical milling machine pictured in Figure 6-1, most machinists today could not even set a shaper up let alone get some work done.

Bridgeport Mills

When everybody thinks of the vertical milling machine, the first thought is a "Bridgeport." This is the machine by which all others are judged. Now I have run a few Bridgeports in my time and I would have to say they are a nice machine. They have the look and feel of a well-made machine. All the levers turn and lock smoothly with just the right click and feel. The height, width, and depth lend themselves to machinists of average height and reach. They are also far from perfect.

What happened to the Bridgeport mill often happens to an average design that meets high demand and sales. The design stays static and all

Figure 6-1: A Bridgeport vertical milling machine.

the design flaws are faithfully reproduced in the army of clones marching out of the factory. Worse yet, the competitors who have a chance to correct the problems instead copy the flaws in their own brand of knockoffs.

I can also say that if I had been the president of Bridgeport, I would probably have done the same thing. Why mess with success? Bridgeport had the manufacturing capacity and most evolved design at the time when there was a high demand for machines.

The only reason I even bring it up is so designers and machinery builders can learn and hopefully advance the state of tool design to the next level. The design should never have become static in the first place. Anybody who has spent time on a Bridgeport or clone of one will be able to relate to the basic design flaws.

The Y-axis dovetail ways are much too narrow in relation to the length of the table. The entire X-axis can be rocked back and forth when the gibs are not set snugly. Set the gibs too tight for minimal play and your arm is dead at the end of the day. The Y-axis ways should be extended to double their current width. This has the added benefit of covering and protecting the exposed ways behind the table where all the chips land and damage the ways. If you have ever looked at a clapped out vertical mill, this is one area where they show their age.

The head tipping feature is grossly offset from the center of gravity of the head. Heavy cuts can easily knock the head out of tram. The pivots should be on the centerlines of the spindle at least. At the very least, eliminate to the front-to-back tilt feature and go with a single right-left tilt.

Suggested Improvements

The entire drawbar assembly could be improved drastically. Think about tool retention in CNC equipment. Thousands of lost man-hours could be mined with a few simple improvements in this area.

Quill locking feature. I have seen quite a few novel methods to keep this limp locking device from dragging. They are all weak band-aide fixes for a bad design.

Axis locking screws. These tend to push the axis off position when activated. How about some blade type locks as per jig borer design?

Acme lead screws. Excellent ball screws have been available for years. When are the manual

Figure 6-2: An indicator with a vertical dial.

machine manufacturers going to take advantage of these low-friction enhancements?

So, until some clever person decides to really take an objective look at the vertical turret milling machine and make some changes, we're all stuck with what we have. Most machinists are pretty clever people themselves. There is more than one way to make one of these machines make parts and money.

Okay, I'm now switching off soapbox mode and on to the manual mill section.

Get yourself an indicator with a vertical dial (Figure 6-2). This beats craning your neck like a bird hunting a worm all around the mill when zeroing a part in.

For tramming the head in, I made several long indicator holder bars to sweep a larger arc (Figures 6-3 and 6-4). If you are having trouble indicating a

Figure 6-3: Indicator holder bars.

Figure 6-4: The longer bar sweeps a larger arc.

Figure 6-6: Using a combination square.

bore that you think should be round, be sure to check your head tram condition. Typically you would see a longer direction or equal but mismatched sweep numbers. If the head is out, you're out on your head.

Button type indicators slide over the tee slots easier than a test indicator when tramming (Figure 6-5).

Use your high-quality combination square on the mill (Figure 6-6). You will be surprised how close you can get. This saves decades of indicating time. I checked one of my combination squares against a master square comparator recently and was happy to find that it was within .002 of square at ten inches off the surface plate.

Figure 6-7 shows special toothed wedges I use as backstops in the mill. I used to use 5/8 dowel pins like everybody else until I found these. These wedge tight in any width tee slot. You can vary the

height easily and even use them for special clamping jobs.

Figure 6-8 shows two wedges set into the tee slots as a backstop for a plate. These unique wedges are called Quoins. They were used in the printing industry to keep type securely into the type case. As usual, I have bent them into service for a different task.

Figure 6-7: Special toothed wedges used as backstops.

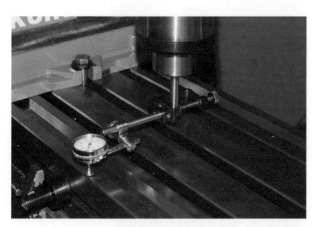

Figure 6-5: Button type indicators.

Figure 6-8: Two wedges set into the tee slots.

Figure 6-9: Cut long shanks off the drill chucks.

Cut the stupid long shanks off your drill chucks (Figure 6-9). The R-8 collet is only an inch long inside. How much shank do you need up there? Think about all the time you spend cranking the knee up and down to get the chuck in and out with that long shank.

For that matter lose the R-8 shank on your drill chuck. If you have a 5/8-diameter straight shank on your drill chuck (Figure 6-9), you will save hundreds of collet changes a year if you buy a few common sized end mills with the same shank size as your drill chuck. That's in addition to all the time wasted cranking the knee.

Knee crank trick

This is an old shop trick to be used on apprentices and newbies. Everybody hates manually cranking the knee up after having near the bottom of the travel. The trick goes like this. As the hapless victim walks nearby, pretend to be listening as you slowly crank the knee up. Give a little harrumph of concern just as they get close. If you're lucky, they will ask what's wrong. If not, call them over. Ask them to crank the knee up and see if they hear the noise. You will need to be non-specific about the actual noise. As they crank it up, listen and comment appropriately. "There it is! Did you hear that? Crank it again, a little faster, that's it. Did you hear it that time?" When you reach the desired height, shake you head in mock concern and say something like, "We need to keep an eye on this machine" or something along those lines.

Buy a couple of drill chucks. If you have different diameter shanks on them, you can save time on tool changes when you have the same shank diameter as your cutting tools.

Don't put end mills in quick-change drill chucks. This is very tempting at times but is a pure rookie move. If the end mill chatters for a billionth of a second, the chuck loosens and all hell breaks loose. I saw somebody do this one time. . . .

Wear out a drawbar once in a while. Drawbars are cheap compared to work that's spoiled because the collet wasn't tightened enough.

Remove the drawbar every so often and put a drop of oil or light assembly lube on the threads. You should be able to spin this with your blistered pinkie finger. If it doesn't spin freely, get a new one.

Use the spindle motor to rapid traverse the collet out once you have it broken loose with the wrench. Only hold the drawbar with your fingers lightly and catch the collet as it falls out. Never use the wrench.

Better yet, invest in a power drawbar (Figure 6-10). If you haven't tried one of these you should. This device has a very short payback period. Unless you use your right angle head on a

Figure 6-10: A power drawbar.

Figure 6-11: Hold your hand on the part.

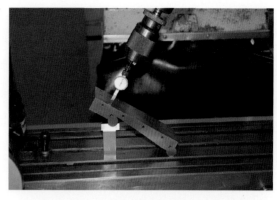

Figure 6-13: Sweeping the face of the bar.

daily basis, the argument about slow changeover doesn't hold any water. Time a few tool changes and do some math to see what I mean. Most people would easily invest in a DRO for a manual mill for the convenience. Once you have tried a power drawbar, you will wonder how you got along without it. No more smashed fingers or wrenches rattling around over your head.

Hold your hand on the part when using tricky or dicey setups (Figure 6-11). Your hand will detect a part shifting before your eye will see it, giving instant feedback to the other hand that's cranking the feed handle.

Get used to using a sine bar. These are simple-to-use, deadly accurate, angle-setting tools. Your sine bar should span across the ways of your standard milling vise. Don't think of them as too precise to use for everyday work. Smaller sine bars are handier for manual mill work. A three-to-five inch center distance is perfect.

Use your sine bar to set head or vise angles (Figure 6-12). You can also sweep the face of the bar just like you would when you tram the head to set a precision angle (Figure 6-13).

Do a sanity check with your protractor to confirm angle settings (Figure 6-14). I saw somebody make a little math boo boo once. . . .

You can use a gage pin (Figure 6-15) or an adjustable parallel (Figure 6-16) to set your sine

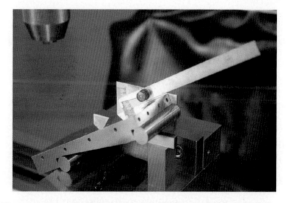

Figure 6-14: Confirming angle settings with a protractor.

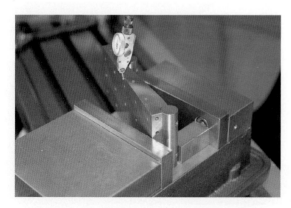

Figure 6-12: Setting the head or vise angles.

Figure 6-15: Using a gage pin.

Figure 6-16: Using an adjustable parallel.

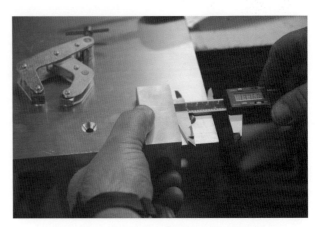

Figure 6-19: Make the block an accurate size.

bar quickly. This can save stack-up math errors with gage blocks. As a good measure, always caliper a stack of blocks to confirm your math (Figure 6-17).

Sometimes it's faster and easier to make a quick drill fixture to drill and tap holes in the edges of large plates(Figure 6-18) . The setup time

and handling can be murderous for just a few holes. Make the block an accurate size so you can locate it precisely on the plate (Figure 6-19).

Figure 6-20 shows a technique that beats setting up the right angle head or hanging the plate off the side of the mill. If you use a hardened drill bushing, the hole comes out straight and in the correct position. You can even make a second fixture for tapping the holes if the additional time is warranted.

Figure 6-21 shows a poor man's knee rapid feed. This will save you quite a bit of time over the course of a year. It has wrench flats on the shank so you can still make fine adjustments with a wrench. Nobody can say I'm **not** a cheapskate.

Once the mill parts are complete, then cut them off with a slitting saw or key-seat cutter (Figures 6-22 and 6-23). This works particularly

Figure 6-17: Confirm math with calipers.

Figure 6-18: Dripping and tapping holes.

Figure 6-20: Another technique for drilling a hole.

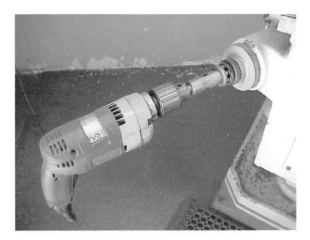

Figure 6-21: A knee rapid feed.

Figure 6-24: Using your right angle head.

well for very small parts. Be sure to keep your eye on the part when it comes off or you might spend more time looking for it than it took to make it.

You can use your right angle head like a precision cold saw to neatly cut parts accurately to length (Figure 6-24). This is a good use of a retractable stop to eliminate parts jamming between the stop and the blade. Use blades with enough thickness to cut straight.

You should be able to cut parts within a couple of thousandths (Figure 6-25). Make sure the bottom of the right angle head clears the part and the vise (Figure 6-26). Also be sure to retract your stop before the part comes off or it may jam.

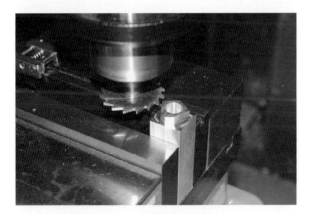

Figure 6-22: Cutting off mill parts.

Figure 6-25: Cutting parts precisely.

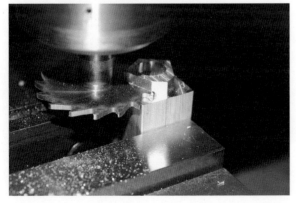

Figure 6-23: This technique works well for very small parts.

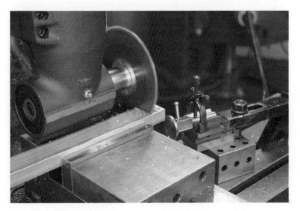

Figure 6-26: Clearing the part and the vise.

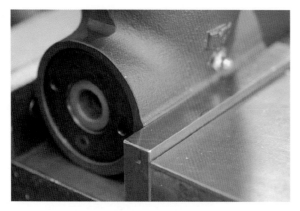

Figure 6-27: Dropping the head into the jaws of the vise.

Figure 6-29: The part is rotated by the rotary table.

You can quickly align the right angle head if your vise is straight with the world. With the clamp screws for the right angle head lightly clamping the body, drop the head into the jaws of the vise and snug the vise slightly (Figure 6-27). For fussy work, you will still want to indicate the head for perfect alignment. By the way, don't trust the flats on the side for fussy work. Indicate a test bar held in a collet.

Cut odd radii with a boring head (Figure 6-28). This is a tube-bending die for an odd centerline radius. The part is rotated by the rotary table with the boring tool cutting edge on the centerline (Figure 6-29).

Cut chamfers with a standard countersink. Stay off the tip for best results.

Place a piece of brown paper under plates clamped to the mill table. This tiny bit of paper acts like a brake lining to keep your part from slipping.

Try using annular cutters (Figure 6-30) for hole making in the mill. They produce accurate holes and need a fraction of the feed pressure to make them cut. You also end up with a neat little slug that you can use for something else (Figure 6-31). Do not stack cut with them, period. Annular cutting tools are much more efficient than drill bits

Figure 6-30: Annular cutters.

Figure 6-28: Cut odd radii with a boring head.

Figure 6-31: Slugs that can be re-used.

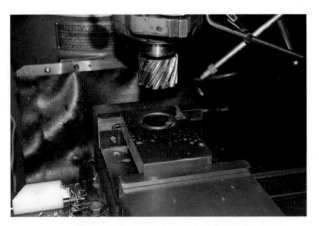

Figure 6-32: Annular cutting tools are efficient.

Figure 6-34: Cleaning the teeth of the hole saw.

(Figure 6-32). The cutting speed of a normal drill bit approaches zero at the tip. Essentially the center portion of a drill bit is broaching it way through the material. Annular cutters maintain a more uniform cutting speed and convert more energy into hole and less energy into chips.

These cutters could not care less if they cut a full hole or some fractional part of one (Figure 6-33). This example shows a two-inch hole in one-inch steel. Hole time was less than one minute, hole accuracy $+/-$ .002. The cutters can stay in the cut because the helical flutes on the outside extract the chips, unlike a holesaw.

When using hole saws in the mill, follow these suggestions for excellent results. Only stay in the cut for two or three seconds at a time. Back the saw out after each three-second peck. After two pecks,

clean the teeth of the hole saw with a small wire brush or air hose while it's running (Figure 6-34). This clears the chips out of the teeth and keeps it cutting. Another thing you can do is pre-drill one-to-four holes in the saw groove to break the chips and strip them out of the teeth of the hole saw (Figure 6-35).

Mill your blocks square with the end cutting surface of your tool. Less tool flex makes for better parts. Learn the 7-step method of squaring blocks without a tool change.

Check flatness of parts with a three-point leveling system (Figures 6-36 and 6-37). Level the underside datum to the same indicator reading. Then sweep the top surface, flip, and repeat. I bet you always wondered what those pointed screw tips were for.

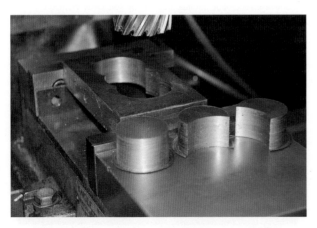

Figure 6-33: Annular cutting tools are versatile.

Figure 6-35: Pre-drilling holes in the saw groove.

Figure 6-36: Checking flatness of parts.

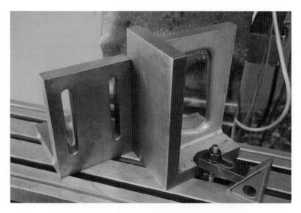

Figure 6-39: An extra tall V-block.

Use a V-block for holding round stuff in the mill vise (Figure 6-38). It gives three-point contact and automatically squares the stock to the jaw accurately.

Extra tall V-blocks can be set up with a couple of angle plates (Figures 6-39, 6-40, and 6-41). One advantage of doing this is you can set up the angle between the two plates to anything you want. Be sure to indicate the faces vertically to insure they are straight with the world. You can also leave a space between the plates for clamping purposes.

Or you can make a soft jaw that has a V-groove already in it for three-point holding a variety of round items (Figures 6-42 and 6-43).

Figure 6-37: Using a three-point leveling system.

Figure 6-40: Adjusting the angle of a V-block.

Figure 6-38: Using a V-block.

Figure 6-41: Another V-block.

Figure 6-42: Three points hold round items.

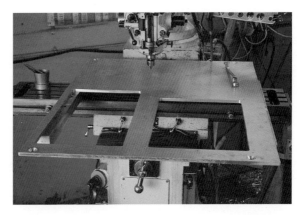

Figure 6-45: Spreader bars help when working on large plates.

For working on large plates, make yourself a couple of spreader bars (Figures 6-44 and 6-45). The spreader bars have counter-bored bolt holes on the same centers as the tee slots. Large plates can be C-clamped to the overhang directly to the bars. Make the spreader bars sacrificial out of something soft.

If you round the cutting corners of a standard drill bit with a honing stone (Figure 6-46), it will leave a better finish in the drilled hole (Figure 6-47). This is a sneaky way to get around an odd reamer size you don't have. You can also hand grind a twist drill a tiny bit off center to get it to cut a whisker larger.

Figure 6-43: A soft jaw with a V-groove.

Figure 6-46: Using a honing stone to round corners.

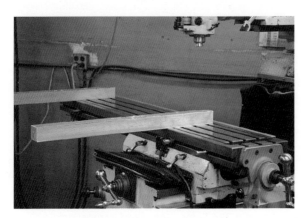

Figure 6-44: Spreader bars.

Figure 6-47: Improving the finish of drilled holes.

Figure 6-48: A slip of paper helps with measurement.

Figure 6-50: More than one setup on a machine.

A simple way to measure the back clearance or relief angle of a hand ground drill bit is to use a little slip of paper (Figure 6-48). Wind the paper around the cylindrical part or the drill body smoothly and mark the point where the corner meets the edge of the paper. The angle formed by these points is the clearance angle if you ever wanted to measure it (Figure 6-49).

With an oversize table on your vertical mill, you can have more than one setup on the machine (Figure 6-50). All our machines are oversize tables; the dovetails are wiped clean all the way to the ends of the travels. We have really made use of that extra work envelope.

You can use castable urethane to support thin-shelled parts for milling or turning (Figures 6-51 and 6-52). This Delrin part would have been

nearly impossible to machine with its .03 wall thickness without internal support. Be sure to use a little automotive wax or mold release so the part pops out.

Figure 6-53 shows another example of using a casting compound to hold or support a part that would be difficult at best to fixture quickly. The tube

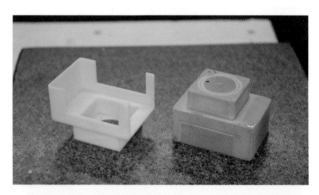

Figure 6-51: Thin-shelled parts for milling or turning.

Figure 6-49: Forming the clearance angle.

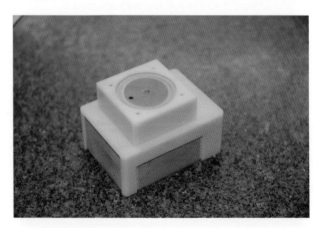

Figure 6-52: Some parts need internal support.

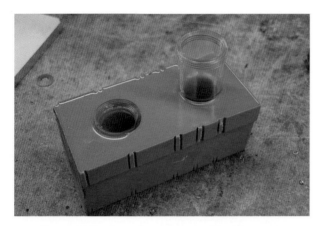

Figure 6-53: A casting compound holds a part.

was held in a chuck in the mill and the casting compound was then poured in an old parts box and around the part to encase it.

The outside of the casting compound was squared up. Then the block was split with a band saw down the parting line we wanted (Figure 6-54).

Now we have a perfect negative shape we can use to register additional parts for machining (Figures 6-55 and 6-56).

This urethane material called Pro-Cast sets up in about 20 minutes, making it a real handy trick for holding impossible parts for precision machining (Figure 6-57).

The next two figures provide another example of using a castable material to support a part for machining. Figure 6-58 shows a low-temperature

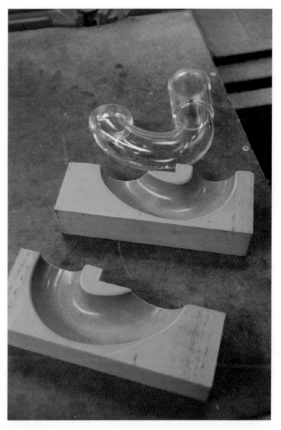

Figure 6-55: Creating a perfect negative shape.

melting metal alloy called "Cerrobend." It melts at 150 degrees F. The part is an aluminum heat sink where the thin fins were filled with the metal and then machined. The wrapping is aluminum foil tape, which forms the dam to contain the liquid metal (Figure 6-59). Extra-melted material can be reclaimed and cast into pucks in a muffin

Figure 6-54: Splitting the block down the parting line.

Figure 6-56: The negative shape can register additional parts.

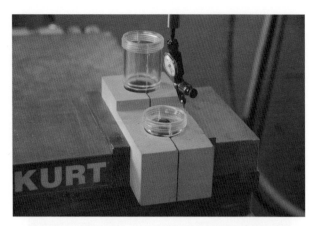

Figure 6-57: Pro-Cast holds impossible parts.

Figure 6-59: Wrapping with aluminum foil tape.

Figure 6-58: Cerrobend is a low-temperature melting metal alloy.

The pins at the bottom are what the part rests on to take the machining thrust (Figure 6-62). Around the back side there is a stiffener under the mill table. A small jacking screw against the dovetail bottom stiffens and squares the assembly (Figure 6-63)

You can use your drill chuck as a hand tapping guide if you leave the collet loose (Figure 6-64). Don't forget to lift it up before you move to the next hole. I saw somebody do that once. . . .

Figure 6-65 shows homemade broaching tools for cutting blind keyways or shaving internal corners square in the milling machine. They cut like a vertical shaper might. The shanks are eccentric so the tool can enter the bore size and keep the cutting

tin for the next use. This material works best if not overheated from its liquid point very much. Oil the part before pouring to make the material completely melt out on removal.

Put a short wrench on the vise to keep apprentices from over-tightening mill setups (Figure 6-60). Use the clock method to describe the handle position. "Okay, only tighten this to three o'clock, got it? Not two-thirty and not three-thirty." This only works if they don't have a digital watch. . . . We add the nylon tie to keep the handle from falling off the clock.

Figure 6-61 shows a handy mill fixture for drilling and tapping the ends of long bars. It bolts to the mill table and the mill head is rotated over to one side. On one side, the fixture has a little vertical fence to make sure the part is square.

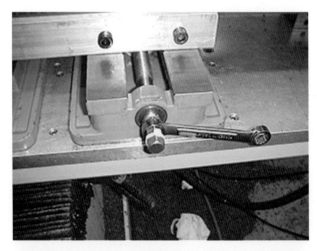

Figure 6-60: Using a short wrench to prevent over-tightening.

Figure 6-61: A handy mill fixture.

Figure 6-63: A small jacking screw.

Get a cheap air ratchet for changing chuck and vise jaws (Figure 6-67). I replace the mounting screws with shorter ones for less time spent spinning. Be sure to take any cutting tools out of the spindle. The air ratchet can buck your hand into the tool if you're not ready for the torque reaction.

edge closer to centerline. Don't tell anybody you can do this or they will want it done all the time. It's kind of like being a dentist and scraping some tartar out of the garlic-eating champion's mouth. This is for special cases only. If you have to cut a blind keyway like this, put a relief at the bottom for the tool to clear the material and break the chip (Figure 6-66).

Figure 6-64: Using the drill chuck as a hand tapping guide.

Figure 6-62: The part rests on the pins at the bottom.

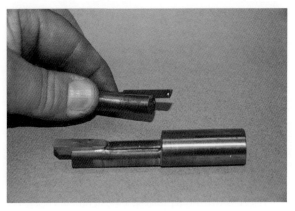

Figure 6-65: Homemade broaching tools.

Figure 6-66: Cutting a blind keyway.

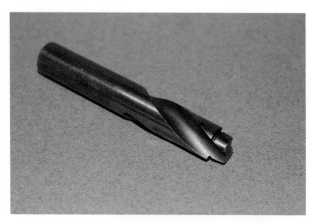

Figure 6-68: An easy-to-make counter-bore.

Figures 6-68 and 6-69 show an easy-to-make counter-bore made from a normal twist drill. It's handy for those odd sizes that come up when the designer doesn't have a regular counter-bore chart. A quick spin on the surface grinder or Deckel tool grinder and off you go.

After tipping the head, return to zero with a square first (Figure 6-70). Tram with an indicator after you have the head close using your square. You will get within the small range of the indicator easily with this quick visual method and save yourself some time indicating.

Try using some soft aluminum filler rod swiped from the welding department to help secure multiple parts. The soft round wire squeezes down and makes up the small differences in part width

Figure 6-69: The counter-bore is made from a twist drill.

(Figure 6-71). This trick works best on hard parts like steel or stainless that you have a fair amount of clamping pressure.

Put the soft wire in vertically for multiple stacked parts (Figure 6-72). The round wire deforms more easily than flat material because the pressure is on a point instead of a line.

Figures 6-73 and 6-74 show another great trick for holding multiple parts securely. This is a standard dovetail o-ring groove cut into the soft jaw face (Figure 6-73). A cutoff piece or o-ring cord stock provides the right amount of squeeze and friction to hold multiple parts securely (Figure 6-74). You can also use different o-ring materials for different types of part holding.

You can insert a sliver of shim stock in one side of a collet to get an end mill to runout and cut a

Figure 6-67: An air ratchet helps change chuck and vise jaws.

Figure 6-70: Returning to zero with a square.

Figure 6-72: Working with multiple stacked parts.

Here is a great way to avoid tipping the head in the milling machine. The tilt plate shown in Figures 6-77 and 6-78 can adjust up to 50 degrees of angle relative to the table surface. It's great for those quick angle jobs that are a hassle to tip the head for. You can set the angle with angle blocks or

whisker bigger. This works when you need an end mill to cut a few tenths bigger.

Figures 6-75 and 6-76 show one of the best things I have ever made. This small 6 × 6 sub-plate with small-sized strap clamps is great for holding down small parts. You can also screw a parallel or a sheetmetal V-plate to the sub-plate as a fence for doing multiple parts.

Figure 6-73: A standard dovetail o-ring.

Figure 6-71: Making up small differences in part width.

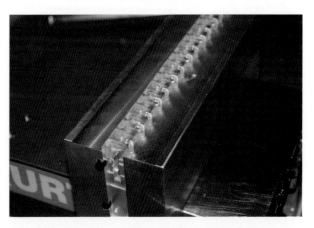

Figure 6-74: Holding multiple parts securely.

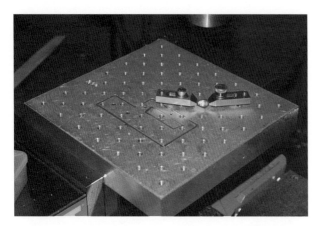

Figure 6-75: This sub-plate holds down small parts.

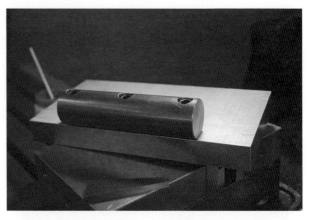

Figure 6-78: The tilt plate is good for quick angle jobs.

use can use an electronic level that was zeroed on the table surface (Figure 6-79). If you really need to you can even do compound angles with this setup.

When setting precision head tipping angles, you can indicate the face of a correctly-set sine bar for those odd angles (Figure 6-81). You can also

use a toolmaker's ball or large ball bearing to accurately align the axis of the spindle with a rotary table axis once the head is tipped (Figure 6-80). These two tricks will come in handy in the next section.

For accurate true position location, drill undersized and then use a single point boring

Figure 6-76: The strap clamps hold the parts.

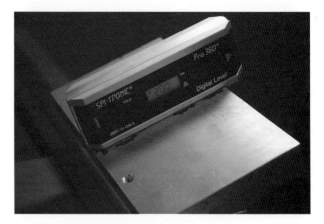

Figure 6-79: Using an electronic level.

Figure 6-77: This tilt plate adjusts up to 50 degrees.

Figure 6-80: Setting precision head tipping angles.

Figure 6-81: Using a toolmaker's ball or large ball bearing.

Figure 6-83: Drilling and following with a plunge.

Figure 6-84: Parallel retainers made from sheet metal.

tool to establish the true centerline of the hole (Figure 6-82). After a light cleanup cut with the boring head, you can finish ream the hole to size or just use the boring head. As an alternative, you can drill and follow with a plunge with an end mill (Figure 6-83).

Figure 6-84 shows a different twist on parallel retainers made from sheetmetal. These leave the

area under the part clear for who knows what to fall into.

Spherical Surface Generation in the Manual Milling Machine

Figure 6-85 shows a unique manual mill method for generating geometrically true spherical surfaces. This technique can be used to machine convex and concave spherical surfaces. The only tools needed other than the milling machine are a boring head and a rotary table, two common mill accessories. If you have a CNC lathe or mill, this is really just an academic exercise. The principle is interesting in that it is self correcting and self proving, which is not true with CNC equipment. If you don't have any CNC equipment, you can add a neat trick to your toolbox.

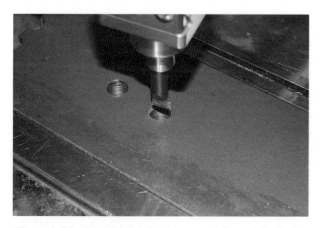

Figure 6-82: Establishing the true centerline of the hole.

Figure 6-85: Generating geometrically true spherical surfaces.

Figure 6-86: Working with convex surfaces.

Figure 6-87: The cutting edge faces inward.

I learned this years ago from my old toolmaker friend Charlie. It's one of those old-timer tricks that I have not seen used anywhere before. When he first told me about it, I was skeptical until I tried it. If you have a computer drafting program, you can make short work of the math and set-up angle. This method is far superior for forming tools and beats the pants off the swinging arc fixtures because the spherical surface is a true geometric generation. The spherical form is limited only by the accuracy of the machine spindle and the rotary table—two intersecting circular paths that produce a true spherical surface.

Imagine a cutting tool that only cuts a hollow circle, kind of like a hole saw. When the cutting tool is set at an angle other than the axis of the rotary table, and the part is rotated under the cutting tool, a spherical surface is generated.

The head is tipped at an angle that represents the chord of the desired spherical segment. A single point cutting tool is used and, depending on whether the form is concave or convex, the cutting edge is reversed. For convex surfaces, the cutting edge faces inward as shown in Figures 6-86 and 6-87. For concave surfaces, the cutting edge faces outward as it would in normal boring head work.

As the cutting tool is advanced into the work, the rotary table is rotated through 360 degrees. The rotary table is also fed into the tool along the x-axis in this case (Figures 6-88 and 6-89). There is an important relationship between the angle of the head and the diameter that the boring head is set.

When you first try this method, I suggest you use plastic so you can quickly see exactly what is happening before you try this on important parts. Don't experiment with hard-to-machine materials when learning, unless you really like turning the rotary table crank. There are three parts you must understand to get controllable results. The first involves the basic calculations. The second is the setup and the third is the execution—actually doing it.

A single-point cutting tool sweeps through a circle that has no thickness on one side of the cutting edge. If you think about how a ring of any size

Figure 6-88: The rotary table is rotated through 360 degrees.

Figure 6-90: The black ring is the cutting edge path.

Figure 6-89: The rotary table is fed along the x-axis.

Figure 6-91: The rings are in full touch with the surface.

smaller than the spherical surface can lay in full contact with the sphere, you can visualize how the cutting action takes place.

The material that projects into the ring is cut away as the part rotates under the cutting tool. This leaves a spherical surface the size of the ring. The black ring is the actual path of the cutting edge (Figure 6-90). You can see how all the rings are in full contact with the surface no matter if the surface is convex or concave (Figure 6-91). Any plane that cuts through sphere produces a true circle no matter what the angle.

In Figure 6-92, we can see the basic graphical setup for cutting a full hemisphere of two inches in diameter. The chord in this case is 1.414 inches. This is the diameter the boring head would be set

at (1.414) or a little larger to cut this diameter. The spindle would be tilted at 45 degrees relative to the rotary table axis to cut a full hemisphere. You can see by inspecting the drawing that no other angle will produce a full half-sphere. The axis of the spindle must be perpendicular to the chord of the segment. The spindle centerline is the midpoint of the chord. The chord is also the hypotenuse of the right triangle that is the maximum rise of the radius and the distance from the centerline to the endpoint of the arc.

For other radii and partial segments, a little math is required to get the chord and the angle. We can use our drawing example to illustrate the math. There is no official name for the diameter that the boring head is set to, so I call it the

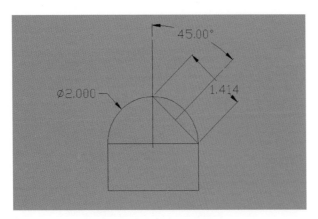

Figure 6-92: Cutting a full hemisphere.

Figure 6-94: Pre-necking the blank.

"Swept Diameter" for our examples, shortened to SD. For OD work, the swept diameter should be set at the chord size or larger. For ID work, the head should be set smaller or the same as the chord.

Angles less than 45 degrees produce less than a full hemisphere. Angles over 45 degrees produce greater and greater portions of the sphere until you reach a maximum of 180 degrees for a full sphere. Once you go past 45 degrees, the boring head must be set accurately to the chord length before you reach the finish diameter. You can adjust this as you rough the part, taking measurements as you go.

We can't in actual practice cut a full sphere in one setup. We still have to hold on to the part and rotate it somehow. In this example, I decided that a .75 diameter stem would be enough to hold on to (Figure 6-93). I pre-necked the blank so the cutting

tool had clearance (Figure 6-94). To produce a full sphere, you must use two separate holding setups. In (Figures 6-95 and 6-96), the spherical section is complete.

Figure 6-97 shows the drawing I used to set up the previous example. The head angle was set at 11.01 degrees using the sine bar method in the

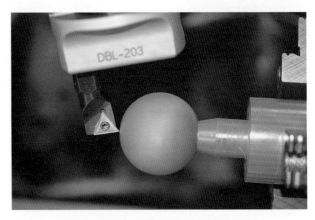

Figure 6-95: Producing a full sphere.

Figure 6-93: Using a .75 diameter stem.

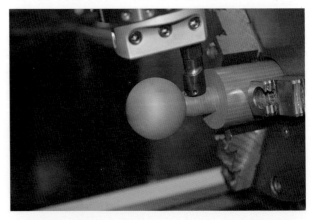

Figure 6-96: The spherical section is complete.

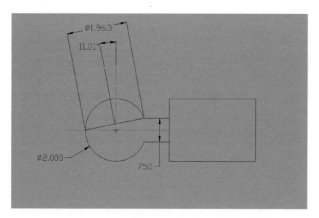

Figure 6-97: The drawing for the previous example.

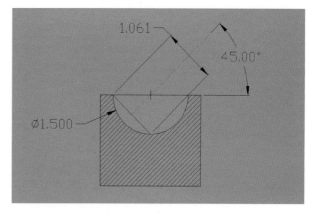

Figure 6-99: Determining the boring head setting.

previous section. I started out with the boring head, cutting a diameter larger than 1.963 inches, then adjusted the boring head as I cut because I was able to take direct diameter measurements off the part.

Figure 6-98 provides an example of a concave radius. The tool edge is pointed inward just as it would be for normal boring work. The swept diameter in this example is the chord 1.061. The head angle would be set to 45 degrees. This produces a full hemispherical cavity.

To calculate other radii and segments, here are the key factors.

$$SD = (D) \sin \theta$$

where

    SD = Swept Diameter (boring head setting)
    D = Desired spherical diameter of part
    Sin θ = Sine of the spindle angle

To determine the boring head setting, SD = (D) Sin θ. In the example in Figure 6-99, our calculation would look like this,

$$(1.50).7071 = 1.061$$

For segments other than full hemispheres, you will have to calculate the chord of the specific desired spherical segment. Note that the head angle setting is important for getting good results. Figure 6-100 shows another example of a large spherical segment.

We are solving for the right triangle that is the rise of the segment and the hypotenuse to the arc endpoint (Figures 6-100 and 6-101). Take a look in the shop math section for a description on calculating the chord length and radii of arcs.

You can also use this trick to cut large radii on the corner of rounds that would be difficult or

Figure 6-98: A concave radius.

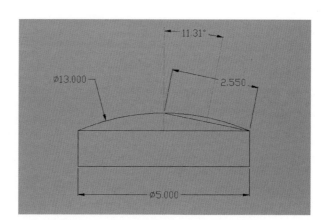

Figure 6-100: A large spherical segment.

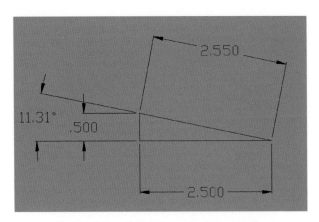

Figure 6-101: Solving for the right triangle.

impossible with form tools. The only requirement for accuracy is that both axes are co-planar in at least one plane. It's best to start the boring head a little large for OD work and a little small for ID work to keep you on the maximum material condition. The boring head can be adjusted on the fly or ahead of time. When I have set the boring head ahead of time, I use a height gage on the surface plate or do a short test cut to adjust the tool precisely. When actually cutting, advance the cutting tool from the outside toward the center of rotation. It makes it easier to see when you are exactly on center without re-cutting the entire surface.

In the next chapter, I will discuss the CNC mill and some of the complex things that can be achieved. In order to fully appreciate what the CNC mill can do, it was necessary to have this chapter on the manual mill directly preceding.

In the age of computer-controlled mills, many young machinists may not have had a chance to operate a manual mill for any length of time. Hopefully some of the things I have illustrated here will give the reader some appreciation what us old timers had to go through to get things done before computers.

CNC Mill

The earliest claimed numerically-controlled machine that I can find was a milling machine unveiled in 1952 at MIT. Much of the early research was sponsored by the military as a technology investment in the future. Like many things, the tools and requirements of the military help push the frontiers of technology by investing large sums of money in areas that may have no immediate use in industry, unlike the structure of private research where the financial payback is on a much shorter leash.

These machines differ from their modern counterparts in that they did not operate directly by computer control, but rather by commands calculated by a computer and read by the machine through a punched tape. Most of these early NC machines were converted from existing machinery as opposed to built from the ground up with the intention of automatic control.

Working with CNC Equipment

Anybody with half a brain can immediately see the usefulness of computer-controlled equipment. The ability of a machine to telescope the work processes so they can run in parallel instead of series can turn a single machinist into a one-person army. Think of it this way: a manual machinist does one man-day's worth of work in one day, right? What if they could do four or six man days of work in that same day? Would you be interested in charging out your hourly shop rate multiple times a day for every machinist? I'm willing to bet you are!

These machines should be thought of as new and faster tools that are available to manufacturing; they allow the company to be more competitive or to increase potential profits. They should not be thought of as machines that displace or obsolete manual machine operators. It's just technology, folks! Every industry, every sport, and every army is looking for a technological edge over the competition.

I tell people that I don't really want to be on the cutting edge of technology, but I sure want to be riding on the blade. Like many of the things I have learned over the years, I was fascinated by these machines and their ability to do the tasks they do with such simple looking instructions. At the time it was all secret code, which I didn't understand. For many machinists the learning curve is steep, especially adding in the complexity of using desktop computers, which now do a large part of the programming duties.

A few words of advice: Don't go the way of the Wisconsin ice cutters! Most experienced non-CNC machinists already know all the hard stuff like speeds and feeds and how to hold things. Most computer jockeys can figure out the controls on a modern CNC machine pretty quickly. The only problem is it takes them ten years to figure out how to hold tools and materials, and then which tool performs what task. The old timers already know that part of the work. So hang in there; take small, steady steps and, by all means, take advantage of these machines.

One challenge I found difficult when I was learning to run CNC equipment was there was absolutely no sense of feel. Manual machinists develop an accurate sense of feel for something like how hard they can push a tool or how deep a cut they can take with a particular setup. This sense of feel is removed with automatic equipment. I had to learn the actual "numbers" for speeds and feeds when programming CNC machines.

The current crop of CNC machinery is so blindingly fast, it is frightening. For most jobs you cannot even approach the maximum available feedrates for this equipment. The world is waiting for the cutting tool industry to catch up with the machine builders. Currently the cutting tools are the gating factors for most metal removal. There are so many factors that are interconnected with the metal removal rate. Material type, work holding, and operation type are just a few that affect how fast material can be removed. Rarely is the machine speed or power the limiting factor.

My experience with CNC machinery is in the jobbing shop environment. Most of the tips and tricks are related to the highly varied work and many different materials found in the jobbing shop. High-volume production-type work is outside my intended scope for the CNC section, but some of the principles and ideas easily transfer to high-volume work.

Try to reduce cycle times for multiple or long-run parts, but never at the expense of consistency or predictability. A machinist or operator can do much more productive and profitable work than babysit a temperamental process.

Work on fixturing and part changing to reduce button-to-button time. This should be the actual measure of a single cycle. If you're just keeping track of how long the machine takes, you're in for a big surprise. On the same note, when you're quoting this type of work, be sure to consider all the factors—not just the programmed cycle time.

Add more comments in the program than you think you need. Memory is not as much of a problem as it used to be. Good programming practice includes comments and specifics about tooling and setup resident in the program.

Try to program in a non-specific machine format. If possible, your programs should be able to run in any equivalently equipped machine. The machine-specific information comes out in the post processor.

Programs that have actually been run and edited at the control are the Holy Grail. Programs downloaded from the machine control should be identified specifically in your program inventory. However your shop keeps track of programs, there should be a simple way of identifying programs that have been run successfully. We add the suffix DL for download when saving a program downloaded from the machine control.

Review the model or electronic information carefully when quoting. There are so many ways the time you have allotted for file conversion or importing can get screwed up. CAM systems work flawlessly on perfect models. Most of the time, the only place to actually get perfect models is at the machine tool shows where they demonstrate the CAM software. Designers find new ways on a daily basis to make the work of the machine shop never boring.

Never forget that at some point in the completion of the job the actual part should be compared to the original, unmolested, customer electronic information received when you started. After all the electronic file conversions, model surface cleanup, hole blanking, surface extending, and other steps we do to a customer's model in your CAM system to get it cut, sometimes it will come back and bite you. Go back to the original source.

If possible, build your programs in such a way that a complete part comes off the machine. There is nothing more annoying than having to re-set up because the part did not pass the first article inspection or all your extra spare parts were consumed in setting up the other operations. Parts move and dimensions drift as more material is removed. If you can get a complete part or parts in hand with each cycle, you are doing well.

CNC Mill

Rather than get specific about types and styles of machines, I think the better way to present this information is in a more general format. Most of what I have to offer has been learned the hard way, by trial and lots of errors. For each type of job or class of work, there are optimal machine configurations and setups. I hope the reader will be able to apply some of my hard won tips and tricks to their specific problems.

Sometimes it's safer and the machine can run longer unattended if you machine interior slugs into chips (Figures 7-1 and 7-2). It takes a little more machine time, but it can be a real part saver and eliminate babysitting or a bunch of M01 (optional stop) lines.

An alternative is to use a large diameter end mill to remove the slug (Figure 7-3). The gap

Figure 7-3: Using a large diameter end mill.

Figure 7-1: Machining interior slugs.

Figure 7-2: Another look at interior slugs.

between the wall and the slug is large in relation to the slug diameter (Figure 7-4). This gap allows the slug plenty of room to jiggle around and fall free without wedging. Make sure there are no obstructions below the part, like a pile of chips to interfere with the slug dropping free.

For big slugs, it's safer to leave a webbing in the bottom and program a M00 (full stop) with an axis retract to give you room to knock out the slug.

Drilling around the perimeter followed by a couple of quick passes with a roughing end mill to within .075 of the bottom was the most efficient way to get this material out (Figure 7-5).

Chip extraction on vertical machines is a big problem. The through hole drilling provided a way for the chips from the rougher to wash out the bottom, and it physically removed a large amount of

Figure 7-4: The gap is large relative to the slug diameter.

Figure 7-5: Ways to get out this material.

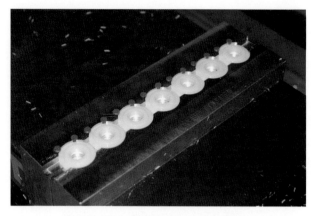

Figure 7-8: Using o-ring cord stock.

the material before the milling operation. We knocked the slug out during a programmed stop in a few seconds (Figures 7-6 and 7-7).

Here is a neat little holding trick. I think everybody has had trouble holding multiple parts in a single vise to take advantage of the speed of the CNC machine. These soft-jaws were cut to hold

the parts with the addition of some pieces of o-ring cord stock (Figure 7-8). The o-ring intrudes into the part pocket by .010, taking up for any size variation in the individual parts (Figure 7-9). Drill the o-ring holes first, then cut the pockets.

Here is another way to use o-rings to help hold parts. Figure 7-10 shows a standard dovetail o-ring

Figure 7-6: Knocking the slug out.

Figure 7-9: The o-ring takes up for size variation.

Figure 7-7: Removing the slug quickly.

Figure 7-10: A standard dovetail o-ring groove.

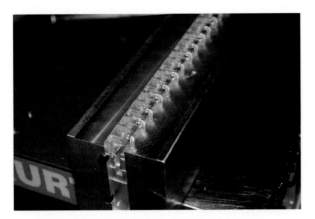

Figure 7-11: Preparing for a facing operation.

Figure 7-13: Preparing for the finish pass.

groove cut into the movable soft-jaw of the vise. The little bit of o-ring sticking out above the surface grips all the parts securely for a facing operation (Figure 7-11). You can use a piece of Delrin rod instead of the o-ring for holding metallic parts.

For one or two parts, you can run a second drill cycle after pocketing partially through drilled holes (Figure 7-12). This will save on subsequent de-burring operations. The vertical burrs on the hole edges are difficult to hand de-burr. If you have the time, you can run around with a smaller end mill and actually radius these edges. For one or two parts, it's hard to justify the additional programming time required.

For some kinds of profiling operations you can leave .010 or less in the bottom of the profile to retain the part for the finish pass (Figure 7-13). I like to leave .030 on the bottom on the pass just

before the finish profile cut. Then on the last pass, I take it down to .010 on the bottom and on size on the profile. The part can then be cut free with a utility knife (Figure 7-14). For this kind of profiling, it is helpful to use bottom-up programming to assure you leave exactly the right amount on the bottom regardless of material variation.

Double check your rapid planes when using bolts or clamps to secure a part for full profiling (Figures 7-15 and 7-16). I think everybody has a little box of bolts next to the machine with the heads machined in some very disturbing ways. Count yourself lucky if this is all the damage you can muster up.

Sandwich thin materials and cut all at once (Figure 7-17). Use a soft cap plate to spread the clamping force over the parts. The parts in Figure 7-18 are .002 thick full hard stainless.

Figure 7-12: Running a second drill cycle.

Figure 7-14: Cutting the part free.

Figure 7-15: Bolts can secure a part for profiling.

Figure 7-18: These parts are .002 thick.

For parts without interior holes to hold them down, you can use a technique I call musical clamps (Figure 7-19). This requires operator intervention at critical times to move clamps and holding fixtures. It helps if your CNC control will allow manual spindle stop with feed hold from the operator station.

A quick custom-sized sub-plate can make full profiling easier by providing clamp access all the way around the edges. Figures 7-20 and 7-21 show a few examples I have made over the years.

The height and generous overhang of the profiling plates in Figure 7-22 make clamping a cinch. Treat these as consumable tools to get your jobs

Figure 7-16: Clamps can also secure a part for profiling.

Figure 7-19: Musical clamps.

Figure 7-17: Cut thin material all at once.

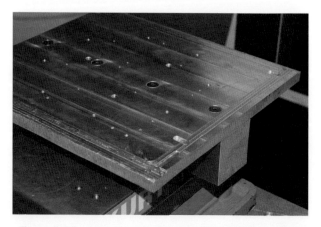

Figure 7-20: A custom-sized sub-plate for full profiling.

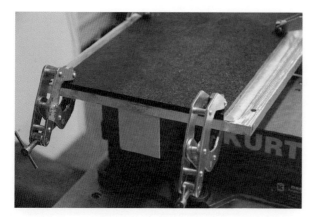

Figure 7-21: This sub-plate provides clamp access.

out the door. Be sure to sink the screws well below the surface so you can take a light skim cut on the top each time you set it up.

The round base allows me to use the handy plates in Figure 7-23 in the lathe as well as the mill. The V-block gives me secure three-point contact clamping.

Figure 7-22: These profiling plates are easy to clamp.

Figure 7-23: These plates can be used in the lathe and the mill.

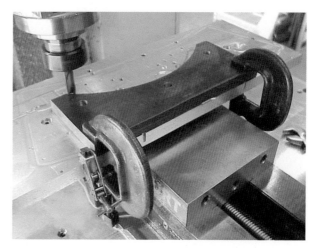

Figure 7-24: Heavy pattern C-clamps.

For a quick setup and secure holding, I sometimes use heavy pattern C-clamps (Figure 7-24). These heavy duty clamps are strong enough to pick up the milling machine so they should hold your work piece without breaking a sweat. This was a quick profiling job in five or six plates. I just clamped them to my aluminum sub-plate that is held by the machine vise and I was off and running. Setup time is all of 2 minutes. I even used a little cantilever clamp as a stock stop.

Be careful when stack cutting. Better to make holes fully into chips than risk a loose slug upsetting your whole job. Use a combination of DOC and a number of passes that leaves a thin web between each of the stack layers. You can also pre-drill out most of the material to make the remaining slugs more flexible.

Cut a shallow recess into the backside of your soft jaws you intend to keep for the future (Figure 7-25). This makes registering the two sides accurately to one another much easier. Another method might be to add a dowel pin hole or vertical slot in the vise jaw. A short dowel pin could be installed into the soft jaws for accurate location. This requires a modification to your vise, however, and has the potential to be cut into when preparing the soft jaws.

Store bought soft jaws are hard to beat from a price point of view (Figure 7-26). They do leave a

Figure 7-25: Cutting a shallow recess.

Figure 7-26: Store-bought soft jaws.

little to be desired in the way of available configurations and features. So all you entrepreneurs out there listen up and get busy! One thing we have done to make the commercial soft jaws more useful is to re-cut the counter-bores and use flathead screws for a little better location repeatability for repeat jobs (Figure 7-27).

Figure 7-28: Important information should be engraved.

Permanently engrave important offset and program information directly into the soft jaws (Figure 7-28). There's nothing like trying to figure this out in six months when the job comes up again. Extra information available at the machine during setup makes for fewer mistakes and faster, more confident setup (Figure 7-29).

Cut a pick-up feature or features directly in your fixtures and soft-jaws (Figure 7-30). There's nothing like a good round hole marked X0 Y0 to set up from the next time you use the fixture.

Figure 7-31 shows an example of using pick-up features so you can easily work outside your machine travels. The dovetail o-ring groove in this large flange weldment was larger in diameter than the Y travel of the machine. It was cut in two halves by flipping the part 180 degrees. An accurate central bore and rectangular feature allowed two super-accurate indicating surfaces to be used

Figure 7-27: Making soft jaws more useful.

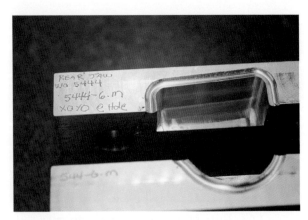

Figure 7-29: Extra information is also helpful.

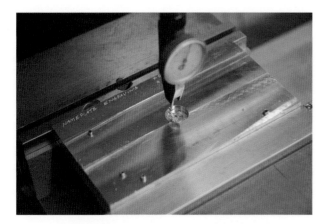

Figure 7-30: Cutting a pick-up feature.

Figure 7-32: Retaining a tool offset.

when the part was turned for the second half. The triangular plate was attached for this specific purpose. This technique can effectively double your machine envelope.

I've never had to do it, yet, but I heard about somebody doing a job that was just a couple of inches outside the X travel of the machine. The job requirements were such that it really wanted to be done in one setup. The solution was to align the part X-axis diagonally with the machine axis. This put the part axis on the hypotenuse of the machine X, Y travels, and gave the person that extra couple of inches they needed to do the job. With modern CAM systems, this is certainly a simple viable option. On the machine in Figure 7-31, that would be an extra five inches of travel. I don't know about you, but there have been times I would have sold my soul for an extra five inches of travel.

A silver metallic felt pen provides a quick, easy way to retain a tool offset directly on a black oxide tool holder (Figure 7-32). It only takes a second to record an offset and it's cheaper than tags. Heck, I can't even buy the nifty little tags for these BT-35 holders, so I improvised like a true cheapskate.

Engraving is super sensitive to cutting depth (Figure 7-33). Your engraving will look lousy if your surface is sloped, dished, or not where you think it is. The best approach is to establish an accurately known Z position just prior to the engraving operation. An alternative is to cut a shallow pocket for the engraving to reside in (Figure 7-34). The pocket also helps protect the engraving from damage because it is recessed. By the way, the pocket technique works wonders when you misspell something.

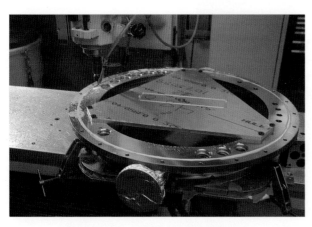

Figure 7-31: Using pick-up features.

Figure 7-33: Methods for engraving.

Figure 7-34: A shallow pocket helps with engraving.

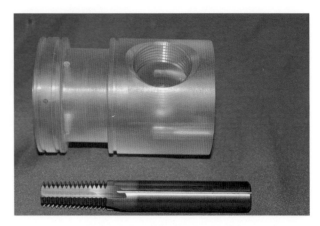

Figure 7-36: 3/4 NPT in acrylic.

You can use an M00 or M01 command just prior to a tapping cycle on difficult materials. Dab a little heavy-duty tapping fluid directly on the tap manually. Water-based coolants leave something to be desired for tough tapping operations. This can make or break a critical job in tough material where a broken tap is not an option. M00 is the preferred method because the default machine condition is a safe one, unlike the optional stop where the operator must remember to have a switch set.

If you haven't tried thread milling, you should. Figure 7-35 shows several different types I use.

Sometimes it's the only way to get a thread in a delicate part. Figure 7-36 shows an example of 3/4 NPT in acrylic. With a large tapered thread so close to the edge, it would have been difficult not to break the part with a normal tapered pipe tap. When thread milling, the cutting forces are low and it's easy to adjust the thread pitch diameter for a perfect gage fit. You can also cut a full engagement pipe thread with less than the full depth of hole you would need, even with a short projection pipe tap.

Don't stress out too much about the ramp in and ramp out moves if you are hand programming

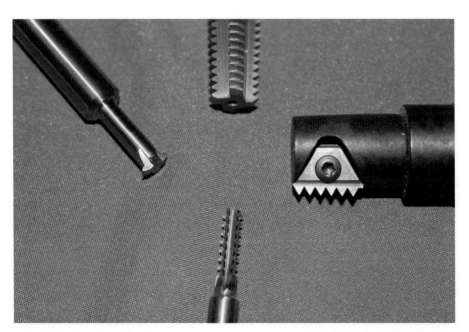

Figure 7-35: Various tools for thread milling.

Figure 7-37: Using a single flute-type thread mill.

thread milling. The straight-in approach works fine and is easier to think about the first few times you try it. Most of the time, you are ramping into air or a thread relief. You can use G41/G42 tool compensation if you like to adjust the diameter, but this can be a little tricky sometimes down the hole of an internal thread. I generally do the entire thread without tool compensation and just adjust the X or Y point where the tool starts to orbit for the helix. I put a marker in the program if I am fussing around adjusting this value so I can find it easily. With the single flute-type thread mill, you can also mill those odd-ball thread pitches that come up every once in a while (Figure 7-37).

I'm sure everybody has experienced the weird phenomena of an end mill or drill lasting for hundreds of parts, replacing the tool "just in case," and then having the new tool only drill or mill one part before failing.

Check a few of your ER-style collet setups for runout if this has happened to you (Figure 7-38). You might be shocked at what you find. These collets are not automatically accurate. If you're having problems with small drill or end mill breakage, check your runout. This is particularly important with small tools that cannot tolerate much runout in proportion to their diameter.

If you find runout, try thoroughly cleaning both the collet and the holder. Then clock the collet in

Figure 7-38: Checking ER-style collet setups.

the holder for minimum runout. The culprit is usually the inside surface of the nut and collets with coolant residue or small burrs on them where the nut touches the collet. Keep these surfaces clean and slippery to minimize problems.

A great way to clean your small collets is in an ultrasonic cleaner (Figure 7-39). Small chips and

Figure 7-39: An ultrasonic cleaner.

Figure 7-40: Plumbing a quick disconnect.

Figure 7-41: This hose can wash out the inside of a machine.

debris worm their way into the collet slits and ruin accuracy. Use a heated solution of degreaser like Simple Green or Omni-All, followed by drying and a dip in a lightweight oil like M-1 or WD-40.

If you are planning your CNC shop, be sure to include a water hose bib near your machines. Machines that operate all day long lose significant coolant water to evaporation. This can be made up with added water or a makeup solution of your water-based coolant to maintain the correct concentration.

Plumb a quick disconnect in line with your coolant line (Figure 7-40). If you make up a short hose with a spray nozzle, it can be used to wash out the inside of the machine after a job (Figure 7-41). One hose fits all your machines. This also makes it easier to steal some coolant for use in another machine in a pinch.

Use your tools for cutting as long as possible to reduce cycle times. You can drill, mill, and chamfer with drill point end mills (Figure 7-42). Sometimes a subtle change in tool selection will allow a tool to be used for several operations, eliminating tool changes and non-cutting time (Figure 7-43). This tip is valid even with modern high-speed tool changes. Vertical CNC machines have less-than-stellar spindle utilization, so anything you can do to keep the tool in the cut is a good thing.

Consider different methods for blank preparation. The blanks in Figure 7-44 were efficiently rough profiled with waterjet cutting. The material

is type 316 stainless. Just because you have a mill doesn't mean you should use it for everything. In this case, the waterjet can fully profile efficiently without any fixturing. Waterjet, plasma, laser, and flame cutting are all processes you should be familiar with and understand their advantages and weaknesses.

Figure 7-42: Drill point end mills.

Figure 7-43: Using a tool for several operations.

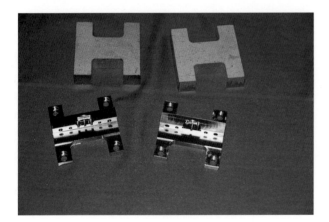

Figure 7-44: Preparing blanks.

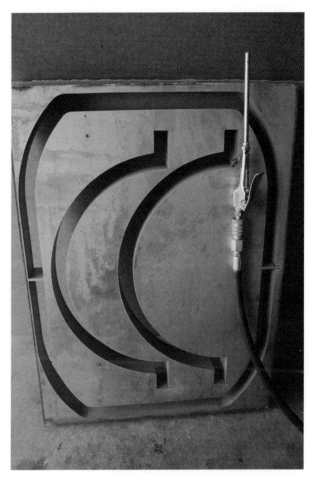

Figure 7-45: The leftover from a waterjet cut.

Use every tool, trick, and option at your disposal to get these jobs out the door as quickly as possible. Speed and momentum have a cost saving advantage when applied to part processing.

Figure 7-45 shows the leftover from a job waterjet cut from two-inch-thick stainless; it's so cool I can't bring myself to scrap it out. The parts were then profiled and finished in the CNC mill. Can you imagine roughing out all the material surrounding these profiles with end mills? The waterjet pierced this material with a neat little hole not much bigger than a 1-mm pencil lead.

Here's a trick for separating ganged parts (Figure 7-46). A 90-degree chamfer mill or drill point end mill is used to follow a contour that will be the parting line between multiple ganged parts (Figure 7-47).

Leave .001–.003 at the bottom for an easy break line. These parts can now be snapped off and easily inserted into a pocketed soft jaw for its second side operations (Figure 7-48). You can see the faint raised trail on the backside of the part in Figure 7-49.

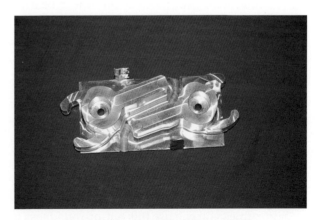

Figure 7-46: Separating ganged parts.

Figure 7-47: Following the contour.

Figure 7-49: The backside shows the break line.

Figure 7-48: Creating a break line.

Circular interpolation is great. However, use a boring head for the roundest possible holes. Single-point boring is still the most accurate for true position and hole roundness. Use single-point boring for precision bearing bores and where critical true position is required, like those found with gear meshes and linear bearing bores.

If you're having problems with hole roundness or true position when circular interpolating, try slowing the feed rate down for the finish pass to even out the quadrant mismatches. Even with modern high speed controls, this little trick will sometimes give you more consistent results. If your plug gage rocks in one direction when you test the hole, this is a sign your holes are not as round as they could be.

Sometimes you can use this out-of-roundness to your advantage. For sliding fits, a few tenths of

out-of-roundness provides an air vent for a blind hole or a place for lubricant to reside.

Keep a machine-dedicated logbook at each of your machines and CAM stations. This logbook gives you a place to record problems and maintenance. It also provides a great place to store all the little quirky bits of information related to each machine—items like, "Check Y-axis limit switch if you get error 1234." If you need to call in outside help, the logbook can be a valuable source of history for the technician. If you work at it and record information like metal removal rates for different tool types, these logbooks can also become a great training aide. This information can become the gold standard of operations for a company, helping at every level from the shop floor to engineering and estimating. On the same note, read the logbooks occasionally and see what kinds of information are being recorded there. If it's worth writing down, then it's probably worth reading.

Document and laminate company setup and operation procedures for each machine. This will pay big returns when training new operators. Keep the length to one operation or task per sheet, then title the sheets. Some examples include *Setting tool offsets*, *Re-starting program at a particular line*, *Adjusting wear offsets*, and *RS232 communications*.

You can attach a short wire and alligator clip to your electronic Z height gage so you can use it on non-conductive materials (Figure 7-50).

Figure 7-50: Using a short wire and alligator clip.

Super glue can be used to fixture parts for machining. The glue used in Figure 7-51, made by Loctite, is called Black Max. This interrupted facing cut is a good illustration of its holding power (Figure 7-52). A quick rap with a screwdriver handle removes the part from the mandrel (Figure 7-53).

Figure 7-53: Removing the part from the mandrel.

For a few prototype parts in the CNC mill, sometimes it's easier to have a big chunk of grabstock. The part in Figure 7-54 was milled in one setup and parted off complete in the manual lathe as a second operation. In this case, adding a generous chunk of grab stock was less painful than figuring how to hold it for the second side,

Figure 7-51: Using super glue to fixture parts.

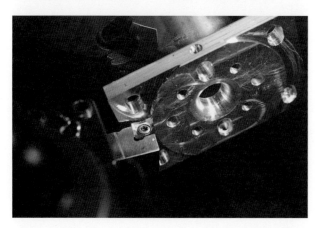

Figure 7-52: The holding power of the glue Black Max.

Figure 7-54: Working with grab stock.

Figure 7-55: Holding small flat parts for full profiling.

Figure 7-57: The wax is coolant proof.

cutting special soft-jaws, and posting multiple programs.

Figure 7-55 shows a special fixture I made to hold small flat parts for full profiling. The dark material in the pocket is a special wax called *Dop Wax*, used by jewelry makers to hold stones for grinding and faceting (Figure 7-56). The entire assembly is placed on a hot plate where the wax melts and wets the blank. After the wax cools, the part can be fully profiled without clamps. It works better than double-sided tape because it's truly coolant proof (Figure 7-57). After completion and a little warming, the part can be removed from the wax. Watch out for excessive heat generation while machining with this method. If the part gets too warm, it will remove itself from the machining area. If you're careful, the plate can be used many

times. The replaceable support pegs are a whisker below the top surface of the plate (–.015). Your part should rest flush on the pegs, slightly embedded in the wax.

An alternate method is to just use the Dop wax on a flat, clean plate. After profiling, the residual cold wax can be broken off the back side and re-used. If you have trouble removing the wax, put the part in the freezer for a few minutes; the wax will pop off even more easily. Any residue can be fully removed with isopropyl alcohol (Figure 7-58).

You never seem to have as many tool totes as you would like. I think the available totes for CNC tool holders are overpriced, which makes stocking up on them painful for the small shop. We made the ones in Figure 7-59 in our sheet metal shop

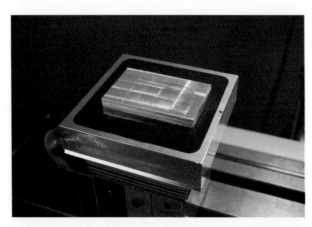

Figure 7-56: *Dop Wax* is used by jewelry makers.

Figure 7-58: Residue can be removed with isopropyl alcohol.

Figure 7-59: A homemade tool tote.

Figure 7-61: Holding the blank.

from 1/8 aluminum. They are more compact than the commercial units and dirt cheap to make. This allows you to "kit" a job's tools ahead of time or even store the exact tools for a repeat job efficiently. I even haul these into the CAM area when I'm programming so I make sure I have everything thought out correctly.

For heavy roughing and ripping, use a dovetailed blank (Figure 7-60). This allows you to use a minimum of grab stock with the highest possible security. The dovetail is a standard 60-degree angle and is .200 in height. Blank preparation is quick and easy. No loose parallels are needed and minimal vise pressure is required for secure holding. You can almost forget to tighten the vise and still hold the blank (Figure 7-61).

These edge grippers, called Talon Grips (Figure 7-62), are a recent addition to my bag of tools. They are heat-treated edge clamps .060 high.

They actually bite into the edge of the blank holding only .060 material. In my semi-abusive testing so far, they have managed to hold everything I've asked them to.

They even come with a nice low profile stop (Figure 7-63). The double-edged grippers are supported deep in a milled groove in the soft jaw

Figure 7-62: Talon Grips.

Figure 7-60: Using a dovetailed blank.

Figure 7-63: The grippers come with a low profile stop.

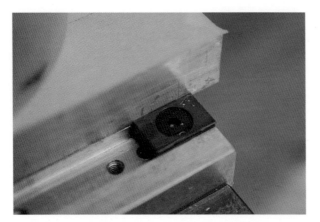

Figure 7-64: The grippers are held down with a #10 screw.

Figure 7-66: Sealing out the coolant.

and held down with a #10 screw (Figure 7-64). This is a great alternative to dovetail prepping mill blanks. No prep is necessary and they hold like Barbie with a Kung-Fu grip.

Double-stick tape works well for some kinds of part holding. Only experience will teach you which kinds of parts can be successfully held this way (Figure 7-65). We use a double-sided tape called "Permacel." It seems to be more coolant resistant than some of the other types we have tried. For a little extra resistance to the coolant, we run a bead around of hot melt glue from one of those cheap hot glue guns you see at the hardware store; this seems to seal out the coolant just long enough to get the job done (Figure 7-66).

Prepare your double-stick base plates carefully for best results. I like to dovetail the base plates and surface them all at the same time for an accurate Z

position like a set of quick change mini pallets (Figure 6-67). Clean the plates with IPA before applying the tape. Position the blank and then go around the blank perimeter with a C-clamp (Figure 6-68) or on the arbor press to seat it firmly into the tape. It's important to press the blank down into

Figure 7-67: Quick change mini-pallets.

Figure 7-65: Double-stick tape works for some applications.

Figure 7-68: Applying a C-clamp.

Figure 7-69: Welding the part to the actual hold downs.

Figure 7-71: Welding tabs from the backside.

the tape. The holding power of double-side tape is all about surface area in contact. You can prepare the blanks offline while the first ones are cutting if you have several base plates. I always like to use fresh tape for each part just to give me the best chance of success. Or depending on how you look at things, a smaller chance of failure. . . .

With certain kinds of profiling, you can actually weld your part to the actual hold downs without the need for extra holding stock (Figures 7-69 and 7-70). Once the profiling is complete, you can remove the hold down bars or tabs and surface the second side. This technique keeps the blank thickness close to the finish size, eliminating large amounts of holding stock that make surfacing the second side an effort.

This method works well for open shapes like "C" and "H" that would be distorted from heavy

clamping pressure. Welding the tabs from the backside also gives us a little chip clearance and drill penetration room under the part (Figure 7-71).

Figure 7-72 shows another way to use dovetails for part holding. These plates are 440C stainless and did not have much extra on the thickness for holding. I wanted to profile and do all the first side stuff in one setup, so we welded a dovetailed plate to the backside of the blank. We used Everdur, which is a low temperature silicon bronze TIG welding rod to attach the plates (Figure 7-73). These were removed after the first side was machined. The low-temperature process did not warp or affect the 440C plate, and the blank was super-secure while flinging smoking blue chips at 500 SFPM with an inserted death mill. The hole in the center was so we could get a little weld in the center of the dovetailed plate.

Figure 7-70: No extra holding stock is needed.

Figure 7-72: Using dovetails for part holding.

Figure 7-73: Attaching the plates with Everdur.

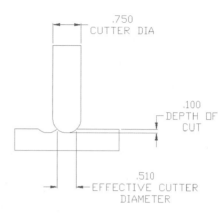

Figure 7-74: Calculating the effective cutting diameter.

All my CNC mills lack rigid tapping, which makes controlling tapping depth a little bit of a challenge. We modified the normal tension and compression holders so they have limited compression length by inserting a longer pin internally; doing so cuts down the compression travel to a third of stock. This approach, combined with a modest spindle RPM, gives us good tap depth accuracy. I always liked the 10 IPM rule for tapping. Take any tap pitch at 10 IPM and just multiply the pitch per inch by 10 to get the RPM. For example, 10-32 tap at 10 IPM gives us $32 \times 10 = 320$ RPM.

Even if you have rigid tapping, you might take a look at a reversing tapping head for your CNC mill. These have the advantage of instantly reversing, whereas in rigid tapping the spindle has to slow and actually stop. If you have huge numbers of tapped holes, this approach has the potential to reduce your cycle time. Torque control is usually adjustable with tapping head, unlike rigid tapping holders.

When doing three-axis contouring with ball end mills, a handy trick is to calculate the effective cutting diameter and increase your rpm and feedrate to take advantage of this offset (Figures 7-74). The full diameter of a ball end mill is rarely fully engaged in fine step profiling and surfacing (Figure 7-75). If the actual end mill diameter is used to calculate the speed, you are most likely running 50% too slow.

Try to de-burr and finish in the machine if possible (Figure 7-76). The added cycle time can be subtracted from the attended machine time. Use small chamfers on unspecified edge breaks whenever possible (Figure 7-77). 45-degree chamfers are more tolerant of part geometry errors than an equal-sized edge break using a radius.

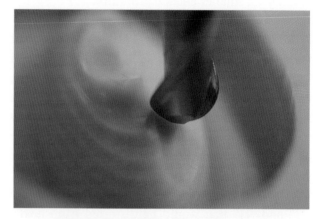

Figure 7-75: Working with a ball end mill.

Figure 7-76: De-burring and finishing in the machine.

Figure 7-77: Using small chamfers.

Figure 7-79: Removing the glue.

Here is a trick I learned from a buddy and have been using with great results. Use hot melt glue to support a part for second operations (Figure 7-78). In this particular example, the floor was very thin. When the back side was surfaced over the thin floor area, there was some chatter, which produced an unacceptable finish. By filling and supporting the thin floor with the hot glue, the second side came out flat with an excellent finish. Thin webs and floors and overhangs can be supported with this cheap easy material. On most smooth machined surfaces, this stuff can be plucked off without leaving residue (Figure 7-79). Use a little spray mold release to make the removal even easier. Denatured alcohol can also be used effectively to help separate the solidified glue from delicate features.

Another trick gleaned from my friend is this interesting use of semi-crash proof soft metal or plastic bed plates on the milling machines. In the shop where my friend works, they prepare the part blanks and secure them to the bed plate with Permacel tape. Plastic locating pins are used to index the part for the second side operations. This technique, used with the hot melt glue filling, allows the un-machinable to be machined with ease. The next set of figures show this technique in action.

In this example, hot melt glue supports an unholdable part for two-side machining. This example is held with Permacel double-side tape to a freshly surfaced sub-plate (Figure 7-80). The first side is faced, then surface contoured to the

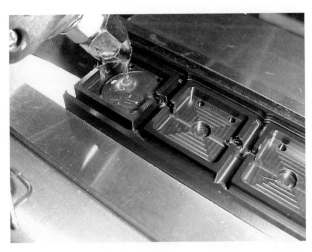

Figure 7-78: Using hot melt glue.

Figure 7-80: Supporting an unholdable part.

Figure 7-81: Facing and contouring the first side.

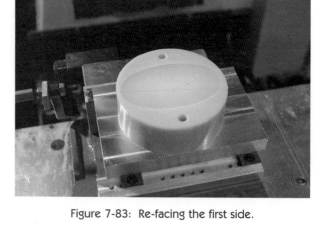

Figure 7-83: Re-facing the first side.

halfway point of the part (Figure 7-81). Be sure to overlap the parting line a little if you are using a ball end mill.

After finishing the first side, the cavity and part are sprayed with a little mold release and filled with hot melt glue (Figure 7-82). If you have a large volume, to fill an inexpensive Teflon-coated pan on a hotplate makes quick work of melting the glue. Locating pin holes are added at this point so the second side is indexed perfectly. Plastic pins are used in case the contouring tool bumps into the pin while the part is cutting. They are only needed when the part is indexed and secured into the double-backed tape. I leave them in because it's easier and I'm paranoid when using double-side tape.

After the glue solidifies, re-face the first side to your Z zero point (Figure 7-83). Use low rpm

to prevent melting the glue. We need a smooth solid surface for the second side so the double-back tape has enough surface area to hold it (Figure 7-84). This method is great for those delicate parts that cannot tolerate any clamping forces without distortion.

In Figure 7-85 we see that the surfacing of the second side has exposed the hot melt glue I filled the first side with. Technically the part is floating in space right now, with no connection to the original blank other than the glue.

You can see though the hot melt glue a little in Figure 7-86. For the next step, I simply band sawed the excess material to get close to the part. Be sure to run the saw slowly because it will re-melt the glue and make a mess of the saw. The last hunk of hot melt was pulled off the part, exposing the finished part.

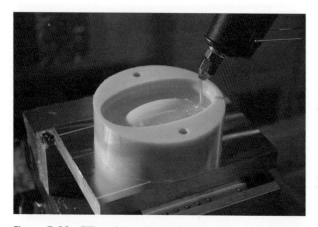

Figure 7-82: Filling the cavity and part with hot melt glue.

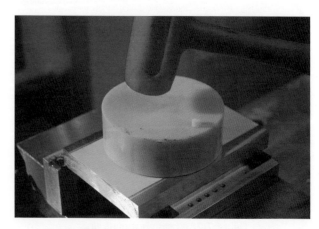

Figure 7-84: Creating a smooth solid surface.

Figure 7-85: Exposing the hot melt glue.

Figure 7-87: A difficult part to hold.

Figure 7-87 shows an example of a part that would be very difficult to hold with normal fixturing methods. For thin shell or other impossible holding jobs, this method can produce fantastic results.

Don't always assume that high-speed steel cutting tools are inefficient on modern CNC equipment (Figures 7-88 and 7-89). On smaller machines, it's typically difficult to take full advantage of the potential of large diameter carbide tooling. On soft materials, like aluminum and plastics, high-speed tools can be run very efficiently at a fraction of the opportunity cost of an equally-sized carbide end mill. For day-in day-out production work on heavy rigid machines, solid carbide or inserted tooling still provide your best tooling cost per finished part cost. We like high–speed, fine-pitch roughing end mills because of the excellent chip control and metal removal rates.

For round parts that have a small diameter variation, a three-jaw chuck mounted to the table is a pretty handy thing (Figure 7-90). Be sure to make some precision spacers to go between the table and the chuck. The chips are difficult to get out of a chuck mounted flat on the table without spacers. Vise soft jaws cut to accept a diameter would put

Figure 7-88: A high-speed cutting tool.

Figure 7-86: Seeing through the glue.

Figure 7-89: Working with high-speed tools.

Figure 7-90: The value of a three-jaw chuck.

Figure 7-91: Record important information.

the true position out by the diameter variation of the part. In this example, we have two diameters with the same end features. Be sure to mount the chuck so you can spin the key 360 degrees.

Clearly mark the machine travels and rated capacities directly on the machine (Figure 7-91). This information seems to come up frequently in the planning meetings and the shop. The more machines you have, the harder it is to remember. In case you're wondering, yes, a few times it's come down to three decimal places.

Off you go now. Good luck and keep the meter pegged.

Figure 7-92

CNC Lathe

When I started in the machining world, I started like many people on the manual lathe. I have a special place in my heart for lathes. In my experience, there is a lopsided ratio of CNC mill and CNC lathe machinists. There are about ten times as many mill CNC machinists as there are CNC lathe people. Why is this? I'm not sure, but my theory is that the lathe is more difficult. I will probably get a flurry of letters telling me how this is not true, but still there remains the difference in the number of machinists who are proficient with CNC lathe operations.

I have heard the argument that the CNC lathe only has two axes; therefore, how can it be so hard? After all, the mill has three axes so it seems more complicated at first glance. Like many things, I have some ideas of my own why there are so many more mill machinists than lathe machinists.

CNC lathes are easier to crash than mills.

Imagine for a minute on your three-axis milling machine that every time you do a tool change the entire tool magazine rapid traverses into the work zone. As it does, it approaches the mill table—which is spinning at 4000 rpm along with your work piece. All the tools hanging down out of the magazine are about to be sheared off by the spinning mass of steel. This would be the mill equivalent to the CNC lathe.

Early on in the history of CNC mills, a tool holding and changing system was adopted that simplifies and helps eliminate much of the danger of spectacular crashes. All the tools are held in individual holders that have a common mounting position and are safely stored well away from the work area. This small factor, I believe, is the major reason why there are so many more CNC mill machinists than lathe machinists; it is so much easier to crash a CNC lathe than a mill because of a basic design difference.

In the standard two-axis CNC lathe, most of the tools are held stationary in a large turret while the work of unknown size and weight revolves to provide the cutting action. Each tool is indexed and brought into the work area with all its companions mounted in the turret. Because most of the tools are stationary, the work holding device which represents a large mass of steel must spin, providing another obstacle that you must plan and account for to avoid a crash when working on a CNC lathe.

If the machine tool builders had really had their thinking caps on, they would have adopted a system similar to the setup used on most CNC milling machines. Only in modern times do we see the trend in some machines, where this kind of system is starting to come out. Sadly, the original decision may have been an economic one for the lathe.

The advantages of the mill system are significant when employed in a lathe format.

In the jobbing shop environment, the CNC lathe can be a real money maker or a real time sinker. It's almost always a tough call to set up and run a small lot of parts on the CNC lathe. These lathes are intended for serious production and often don't lend themselves well to short runs or lots of changeover.

So what is the lower end job size for setting up and running in the CNC lathe versus running it manually? Some of what drives this decision is the part geometry and quantity, but more often than not, it's the setup and programming time. Now think about it this way: If the CNC lathe was as easy as the manual lathe to set up and program quickly, it would reduce the threshold job size to make the decision to run it in the CNC. So we should focus on ways to reduce the amount of setup and programming necessary to get a job out of the starting gate and into the CNC lathe. Keep in mind that the thinking is completely different on a production job where the setup and programming is amortized over a huge number of parts.

Some of the decisions to think about when you run a small quantity of parts on the CNC lathe are:

Is it likely this job will come back? If it comes back, will it be the same? Can we run extra parts and store them for future orders? Do we already run a similar part that we can leverage our tooling or setup with? Is the part geometry difficult enough to warrant the setup and programming effort? Is there any risk in secondary operations downstream that would benefit from extra or setup parts? Can the operator do secondary operations on the parts while the parts run unattended in the CNC lathe to telescope the schedule and manpower used? And, finally the most important, will the customer pay for it?

Any of these reasons are good ones to spend any extra effort required to get the job more automated and potentially more profitable.

CNC Lathe Programming

The typical CNC lathe program is quite short compared to some of the mammoth CAM generated programs seen in milling machines. If you're lucky enough to have a modern lathe with a large memory and a CAM package to program your nice lathe, then some of these programming tips and tricks may not mean much.

Since I am a lathe CAMLESS twit, I have set up my program library into families of parts. I use a couple of generic starting templates that get the machine into a basic configuration to start adding the program elements and events.

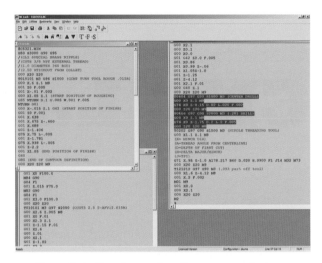

Figure 8-1: Copying and pasting among windows.

Using multiple windows in the text editor, I can copy and paste from one program window to the next (Figure 8-1). Once you have a little inventory of working tested programs in a few materials, it becomes easy to leverage this work against another job. In the world of the low-volume jobbing shop, anything you can do to speed up the non-part-making portion of a process is worth trying.

Copious notes within the program help speed the identification of specific elements that you may want to re-use in another program. This is a great use of the lag time while running parts waiting for the machine. You can make up several generic program scenarios that come up frequently that just need a little quick editing to run.

Use a full-scale drawing to check tool clearances with the actual tools you plan on using (Figure 8-2).

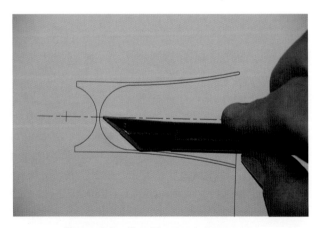

Figure 8-2: Checking tool clearances.

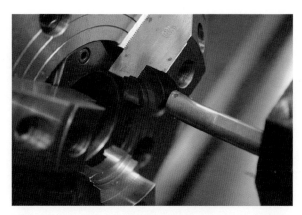

Figure 8-3: Cutting multiple diameters and recesses.

Figure 8-5: Creating soft jaws.

"When in doubt, check it out!" Down in the bottom of a deep bore is not the time you want to find out the tool didn't have enough clearance, Clarence!

For many parts, it's really nice to have them come off the machine complete. It's not always possible or practical to do this. One trick for helping is to cut multiple diameters and recesses into your soft-jaws (Figure 8-3) so you can flip a part around after a pause and finish it in one cycle (Figure 8-4). There's nothing like having the machine set up to run complete parts when the customer calls and increases the quantity. Programming and setting up to complete parts in one setup is a way of leveraging the initial setup to save future setup time on repeat jobs.

Don't forget about your milling machine when you are creating soft jaws for the lathe (Figure 8-5). Inserts and special pocket jaws can be made easily in the mill. The insert method is an easy way to

utilize standard chucks and collets to hold some really weird shapes. In Figure 8-6, simple bored soft jaws hold the milled insert for precise eccentric turning (Figure 8-7).

It's always nice if you can complete a part from one side. One of the things that keeps you from doing this easily is the final little chamfer or edge

Figure 8-6: Using simple bored soft jaws.

Figure 8-4: Finishing a part in one cycle.

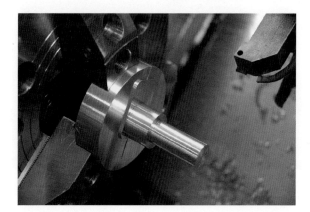

Figure 8-7: Holding the milled insert.

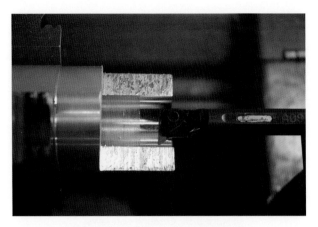

Figure 8-8: Using a threading tool.

Figure 8-10: Pre-chamfering the ID.

break on the inside diameter of a bore that you are parting off into.

You can use a threading tool modified to 45 degrees to pre-chamfer the ID where the parting tool will break through (Figures 8-8, 8-9, and 8-10). You can also profile the chamfer with a standard internal grooving tool. If you choose that route, be sure to pre-groove straight in a few-thousandths deeper than the chamfer before you profile with a regular grooving tool. It can completely eliminate a second operation and the associated handling and price of one extra tool change.

Delicate parts that you don't want going through the chip conveyor can be caught using a magnetic catcher snapped onto the parting station (Figure 8-11). Be sure to put an M01 optional stop

or M00 full stop to remove the parts before too many stack up and cause a problem.

Keep a simple, easy-to-edit program in the control for boring soft jaws or any other repetitive operation that your shop encounters (Figure 8-12). This is a very common setup event in the CNC lathe. If your control supports parametric programming, all the better. A soft jaw setup might be as simple as changing a couple of values at the beginning of the program and touching off the Z axis.

If you don't have a parts catcher on your lathe, use CSS (constant cutting speed) when parting right up to where the piece almost comes off. I then switch to a slower constant speed at this point to do the final parting off. This prevents flinging the part and damaging it against the turret or enclosure.

Figure 8-9: The tool is modified to 45 degrees.

Figure 8-11: A magnetic catcher for delicate parts.

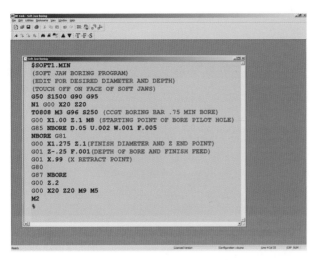

Figure 8-12: Maintain a program for repetitive operations.

```
$SOFT1.MIN
(SOFT JAW BORING PROGRAM)
(EDIT FOR DESIRED DIAMETER AND DEPTH)
(TOUCH OFF ON FACE OF SOFT JAWS)
G50 S1500 G90 G95
N1 G00 X20 Z20
T0808 M3 G96 S250 (CCGT BORING BAR .75 MIN BORE)
G00 X1.00 Z.1 M8 (STARTING POINT OF BORE PILOT HOLE)
G85 NBORE D.05 U.002 W.001 F.005
NBORE G81
G00 X1.275 Z.1 (FINISH DIAMETER AND Z END POINT)
G01 Z-.25 F.001 (DEPTH OF BORE AND FINISH FEED)
G01 X.99 (X RETRACT POINT)
G80
G87 NBORE
G00 Z.2
G00 X20 Z20 M9 M5
M2
%
```

Figure 8-14: A device for catching parts.

Figure 8-15: Turning off the CSS.

If you really need to catch the part, use an M01 or M00 right before the part separates and use a catch cup on a stick (Figure 8-13). Be sure to turn the coolant off in the program so you don't take a bath when you fire it back up. Don't leave very much to part off with the coolant turned off—just enough to give you time to sneak your catch cup into position will do it. The one I made has a telescoping handle off a rolling tape measure (Figure 8-14).

Try turning off the CSS when doing tricky plastics (Figure 8-15). Having control over the cutting speed manually sometimes gives you better control of the stringy chips. Chip control in the CNC lathe on some kinds of plastics can be a real roadblock to unattended work. You can increase the chip load by slowing the spindle or by increasing the feedrate to thicken the chip to get it to do what you want. Sometimes, even a simple task like turning off the coolant can help with chip control in situations like this.

I'm sure everybody has had a problem setting the fine serration top jaws in a CNC lathe power chuck. The 1-mm serrations are easy to set one tooth off and screw up the centering on a blank. Three jaws are difficult to measure and set to the middle of the jaw travel. I made some special labels that have the basic sizes engraved on them backward so I can quickly set the jaws to a diameter range and hit all the serrations on the same

Figure 813: Catching a part.

Figure 8-16: Special labels for top jaws.

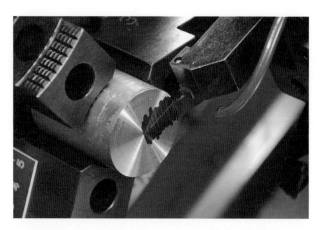

Figure 8-18: Scratching a line along the turret axis.

radius (Figure 8-16). Cost: $5.00 and a little time to engrave on the mill. Be sure to super-clean the jaws where these labels stick on. Make it a habit to brush out the fine jaw serrations with a little scratch brush when you re-install your top jaws.

Sometimes it's hard to align a boring bar with the machine axis and centerline when the manufacturer doesn't provide flats on the tool. To get around this problem, I face the end of a setup bar and darken it with a Sharpie (Figure 8-17). Using a turning tool, I scratch a line a tenth or two deep along the turret axis (Figure 8-18). This line is pretty accurately on center and aligned with the turret X-axis angle. It allows me to twist the little boring bar around and have something for lining up the cutting edge. I use a mirror so I can see upside down to align the tool with the scribed centerline (Figure 8-19).

Once you have set up and used a tool, do yourself a favor. When you remove it from the machine, record at least the X offset on the tube or directly on the tool (Figure 8-20). This will speed the setup of that tool the next time you need to use it. The offset won't be perfect, but it can be darn close. If you're wondering, yes that is a picture of a lathe boring bar in an end mill holder. Sorry, you will just have to keep reading to see this idea in action.

For all my external turning tools, I added a back stop screw that butts against the turret (Figure 8-21). This allows me to retain an accurate X offset if I take the holder out of the machine (Figure 8-22). This pays big dividends during the next setup of that tool because I retain the X offset.

If you have enough machine X travel and your control will allow it, you can run the spindle in

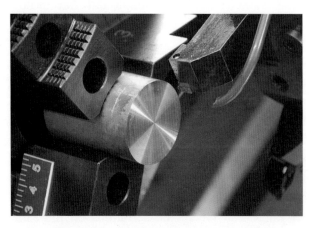

Figure 8-17: Facing the end of a setup bar.

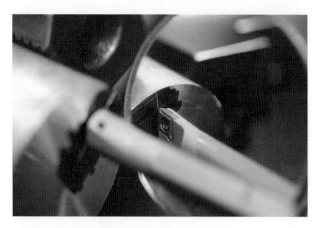

Figure 8-19: Aligning the tool.

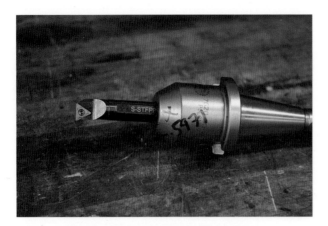

Figure 8-20: Recording the X offset.

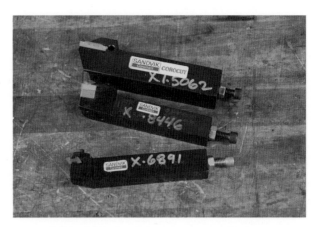

Figure 8-22: Retaining an accurate X offset.

reverse and take a light test cut on the OD of a part, with an internal tool to set an accurate X offset (Figure 8-23). You can set the offset without having to make and measure a bore. On my machine, I enter the offset as a negative X value because the tool is into the negative X quadrant. Be sure to check your control to see if it will accept a negative value and still do the offset math correctly. It's almost always easier to measure an outside diameter accurately than a bore or hole.

For bar pulling operations, it's better to have a spindle liner that is close to your raw stock size. We made up simple spindle liners that drop right in the headstock (Figure 8-24). They use common off-the-shelf pipe and tubing sizes. Have the water-jet or laser cutter zip you out a bunch of disc blanks next time you have less than a minimum order of

cutting. Toss them on the shelf; the next time you need a special liner diameter, half the work is already done. Chop off a piece of tube and weld the discs in place.

A couple of quick little welds (Figure 8-25) and you now have an inexpensive spindle liner for that oddball size you need to run.

Figure 8-23: A light test cut on the OD.

Figure 8-21: A back stop screw.

Figure 8-24: A simple spindle liner.

Figure 8-25: A few welds finish the liner.

Figure 8-27: Using the tool to turn this aluminum hexagon.

Here is a handy trick for cutting diameters and parts outside the maximum turning diameter of your machine. With a special custom shop-made tool, we increased our turning capacity from nine inches to over fifteen inches.

Figure 8-26 shows a close up of the turning tool. We cut the end off a plain old CCGX insert tool holder and welded it to a cold rolled steel shank. Be sure to tack weld it first and then check the machine movement for your clearances and travels before you weld it solidly. This setup puts the tool tip for an OD tool much farther back than you can get with normal OD tools.

Using the tool shown in Figure 8-26, we were able to work well outside our normal turning envelope. The aluminum hexagon in Figure 8-27 was turned using this special tool.

Figure 8-28 shows a method we use to proof programs and even verify dimensions. We make up several test blanks from MDF board laminated together with glue to get any size blank needed. It's cheap and easy to work with. MDF stands for medium density fiber; it is used as the base for cabinet work using plastic laminate. It is quite accurate in thickness and, more importantly, flatness.

The band sawed laminations are glued together with white carpenter's glue and clamped. These are cheaper and easier to use than huge blocks of wax or expensive butter board. It has no grain so it cuts better than solid wood or plywood.

The Fly

During the summer it gets pretty hot out in the shop. Like many machine shops, ours started out

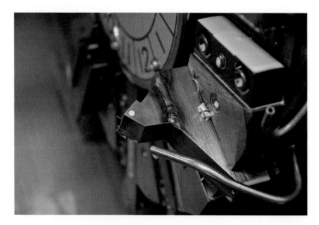

Figure 8-26: The turning tool.

Figure 8-28: Verifying dimensions.

life as a warehouse, so air conditioning is a distant want. With sweat running down my semi-bald head, I was trying to get a part setup in the CNC lathe. This particular part was giving me some trouble because of a tiny little tool compensation problem. I was hot and frustrated.

To make matters worse, a big, fat, bluebottle fly was buzzing around my sweaty head. Periodically the little carrion-eating pest would land in a droplet of sweat that still happened to be connected to my head. With maddening regularity, it would do its little bird bath dance in the sweat droplet and flee as soon as I tried to swipe at it with a dirty paw. This went on for a while as I worked on my programming problem in the heat.

Now I consider myself a pretty patient person, but this was bordering on a declaration of open hostility. Each time the fly landed on my head, it was like the programming gods were sending some kind of message to me reminding me of my flaws as a programmer over and over again. I wonder what hell looks like for CNC programmers? My best guess would be it's an endless string of cryptic error messages followed by eternal heart-rending machine crashes or near misses—while in an elevated temperature environment with a fly thrown in for good measure.

The next hour or so felt like an Inquisition conducted by an expert interrogator. Eventually, however, my feelings went from anger to some kind of resignation to the whole process. Well, after a couple of hours of fighting this fly, I guess I reached the second stage of my downward spiral. I think it was after I wiped my head off to keep the diabolical little creep from landing that I reached that point. I just gave up and let him creep around on my scalp, licking up the salt deposits with his hairy little suction tube.

I have experienced moments of almost divine understanding sometimes when I mentally gave up but kept plugging away in spite of my mental state. This is what happened. I finally figured out what the problem with the program was. I felt a rush of mental relief at solving the problem and a definite surge of the previous loathing for my little winged nuisance.

As always, I tested the program and was finally ready to run it in earnest on real parts. I had run a couple and tweaked the wear offsets, so everything was in order to run. On this part I had set up a bar puller and part counter. It was a small part so I could run over 100 pcs unattended. I opened the enclosure one last time to double check. . . .

The fly flew inside the lathe enclosure.

Blam! The wind from me slamming the door closed probably killed him, but let me tell you folks it was pure pleasure to push "cycle start" that time. I was not going to open that door unless the collet nose flew off and did it for me. I imagined it was like going through a car wash in a convertible with the top down, except the car wash was spewing a wing wetting mixture of oil, water, and flying shards of metal. I think I might just have the tiniest bit of understanding how it feels to be tied to a chair in a dank dungeon being asked for the millionth time,

"Did you check the nose radius compensation? We know you didn't. Why don't you make it easy on yourself and just tell us. This is hurting us more than it's hurting you."

At that moment when you are ready to throw in the towel and say anything to make it stop, three Green Berets kick the door in,

"G41 to you dirt bag," pfhht pfhht echoes off the stone walls of your soon-to-be-vacated hell.

It was that satisfying.

Figure 8-29 shows another lathe quick-change tooling idea. The commercially-available, quick-change tooling systems are extremely expensive and don't lend themselves to highly varied jobbing shop work. I had several special bushings made that match the taper of cheap and readily-available end mill holders. These bushings have the required accurately-ground tapers and locating keys so the tool indexes accurately. The holder in the picture is an NMTB holder, but almost any taper would work. I like the NMTB because they have a standard coarse thread for the retention

Figure 8-29: Another lathe quick-change tooling idea.

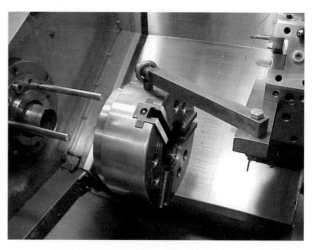

Figure 8-31: Reducing changeover time.

end. I just put a bolt in the back and I can switch between tools in seconds rather than minutes or hours.

Figure 8-30 shows an ID grooving tool held in my quick change system using a standard NMTB end-mill tool holder. This quick change setup is generally for ID tools, which typically are more time intensive to set up in the CNC lathe.

Changing the power chuck on a large CNC lathe can be a real hassle. I made a special tool for helping get the changeover time from chuck to the collet closer down to a half hour instead of an hour or more (Figure 8-31). This pivoting tool clamps in an unused tool position in the turret and the turret is jogged around to help remove or re-position the chuck. Once the chuck is off the hydraulic

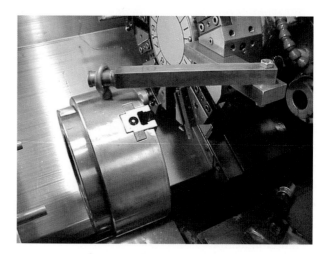

Figure 8-32: Swinging the chuck out the door.

drawtube, the chuck can be swung out the door (Figure 8-32). In this figure, the chuck is going on the machine. I use two guide pins that fit the retaining bolt hole minor diameter to guide the chuck into position and alignment with the draw-tube (Figure 8-33)

Spray some WD-40 on the inside of the window of your CNC lathe (Figure 8-34). This helps keep the window clear so you can see when you're test-ing a program. I tried pretty hard to get a good pic-ture of before and after spraying the WD-40 and failed miserably. Anyone who has tried to see through the shower of coolant, desperately trying to see while testing a new program, can appreciate

Figure 8-30: Holding an ID grooving tool.

Figure 8-33: Guiding the chuck into position.

Figure 8-34: Using WD-40.

this one. The WD-40 helps keep the window a little clearer while you test the program with your finger poised over the cycle stop button.

So, in closing this chapter, you will have noticed that many of these tips and tricks are related to setup reduction. For the highly varied work that our shop sees, this is the roadblock to using the CNC lathe for every job. There is a breakeven point with the set up that is dependent on the part and the quantity of customer orders. I would love to run all the lathe work through the CNC lathe, but it is just not possible, at least for our shop and workload.

All I can say is that it is a fantastic tool, but I'm glad it's paid for and does not have to have its spindle turning all day long to pay for itself.

The Welding Shop

My first experience in the welding shop was when my dad taught me to stick weld in the basement of our childhood house. I was around nine years old and he came home with a brand new welding hood one day. I can remember asking him who the hood was for and being mortified when he told me it was for me because he was going to teach me how to weld.

I donned the nine-hundred-pound, moldering leather welding jacket—complete with the stale sweaty leather smell they all have—and clamped the claustrophobic lightless hood over my head and absolutely loved it. The funny part is that I still use some of the same teaching techniques my dad used on me with the welding wannabes I end up teaching.

I can't remember what my first project was with welding but I know I got into trouble because I went through a 50-lb box of rod just running beads on a flat plate. Something about the arc and the delight of chipping off the slag to expose the gleaming weld underneath caught my interest. The plate started out 1/4 thick and ended up something like six inches thick when I was done. My friends and I used to take turns watching the electric meter speed up when the arc was struck. I wish I could see the electric bill for that month. . . .

I want to define what I call the welding shop (Figure 9-1). My feeling is that the welding shop encompasses all the aspects of putting ideas and products together that happen to be made out of metal. For the most part, this includes anywhere from minimal to massive amounts of welding—all the way from the delicate welding involved with thin sheetmetal up to heavy plate and structural work. Welding is just another tool used by metal fabricators to achieve their goals.

Figure 9-1: The welding shop.

In addition to welding, many other fabrication operations go on in the typical steel shop. from layout to shearing, forming, rolling, punching, cutting, and quite a few others. In my mind, an accomplished metal fabricator has the ability to function in almost any area in the modern metal fabrication shop. A slightly more accurate title for this chapter might be metal fabrication, although this description in itself does not do the trade justice. Most people can relate to the welding shop the best.

You will quickly find yourself going the way of the Wisconsin ice cutters if all you focus on is pure welding. As far as I know, they don't have robots yet that can run every machine in the shop, remove a broken bolt, and drive the truck to deliver the finished goods. Your best bet is always versatility. I have had the good luck of working in shops where everybody did everything and some did certain things better than the rest. I think pure welding is pretty boring. When you add the challenges of fitup, distortion control, and layout, it becomes a much more interesting field.

Getting Started

As dumb as it may sound, the first step in any fabrication job is to do a little planning. This step can be as simple as looking over the drawings or job information and making a cutting list or as complex as designing fixtures or special tools to get the job done. A little pre-thinking goes a long way.

Always take a minute to look at the whole drawing. Be sure to look at any notes on additional processing, special materials, or anything that impacts your part of the process.

Examine the tolerances and understand the job requirements before you cut anything. This is particularly important when geometric tolerances are used. Check for difficult geometric relationships such as projection zones and complex feature relations. The larger the part, the more important it is that you can never assume the designer or engineer has covered all the bases to make your

life easier. In fact, most of the time, you can count on the opposite.

If you need additional information to clarify a job, be sure to get your questions to the appropriate people quickly. It takes a little time to get specific questions answered, so ask early. Written is best. Include a sketch or a photo copy of the relevant section of the drawing.

Think in sub assemblies on complex jobs. It's almost always easier to build and control a smaller part of a large project. It's much simpler to correct a small problem on a sub-assembly than to try to correct that same problem when it is buried and locked in the basement of a large welded fabrication. Sorry, boss, I welded the keel on backward on this aircraft carrier. . . .

Speak up if you see something that you think is a mistake or is going to be a problem. This is not an excuse to complain about every little thing you perceive is hard. If you whine and cry wolf all the time, then nobody will ever listen to you when there really is a problem. You would be surprised how many times specific details that will be a fabrication problem are easily changed by the engineer or the designer. Always remember that they are looking through a different set of glasses. What looks obvious to you as a problem may have been the first solution that was thought of to any problem, and no additional thought was given as they moved on to the next hundred decisions. It's very easy to ask questions, and many times you get the answer you need. I've worked on both sides of the office wall so what seems like a dumb move out in the shop may be the best solution to the ten other problems hinging on that item.

Be realistic about the tolerances that can be achieved with welded assemblies. Don't make the mistake of trying to avoid expensive machine work by substituting even more expensive precision fabrication and the inherent problems that go along with it.

Use these as general tolerance guidelines shown in Figure 9-2 for welded assemblies for

Weldment Size			Minimal Welding		Moderate welding Tolerance		Heavy Duty Welding	
			Close	Normal	Close	Normal	Close	Normal
0-12 Inches			+/- .015	+/- .030	+/- .030	+/- .060	+/- .060	+/- .125
12- 36 inches			+/- .030	+/- .060	+/- .060	+/- .125	+/- .125	+/- .250
36-96 inches			+/- .060	+/- .125	+/- .125	+/- .250	+/- .250	+/- .500

Figure 9-2: General tolerance guidelines.

achievable and consistent results. Trying to force tighter tolerances into the welding department just results in delays and cost increases.

Layout Work

All the lines, arcs, and angles we mark on our work to guide us fall under the heading of layout work. Sometimes work is produced directly from the layout and other times it's used as a reference so we don't make a mistake.

It's tricky deciding how much layout work needs to happen for any particular job. Some jobs require zero layout whereas others are so complicated that a large amount is necessary in order to avoid errors.

Center punching. Grind the tips of your center punch to ninety degrees included angle (Figure 9-3).

This angle is extremely durable. Use a soft metal hammer like copper or soft steel to strike your center punch (Figure 9-4). This keeps the hammer from accidentally slipping off the head of the center punch and spoiling the work when you give the punch a solid blast. Hard steel on hard steel is like a banana peel on an icy sidewalk. Keep the tip nice and crisp so you can feel it click into the scribe lines.

Dividers. You will need at least two sizes of dividers at a minimum. A set with a 6-inch radius ability will carry you a long way. I like a set in which you can interchange a pencil with one of the legs. This is great for template work and gives you a reason to steal those lame short pencils from the library or the golf course. For larger fabrications and arc layout, trammel points that clamp to common bar stock sizes are excellent.

Figure 9-3: Grinding the tips to the center punch.

Figure 9-4: Striking the center punch.

Figure 9-5: Trammel points.

Trammel points that can fit a wide range of bar sizes are superior. The eccentric point type shown in Figure 9-5 will also hold a wooden pencil for paper pattern layout. Fine adjustments are made easily by rotating the eccentric point.

Deep scribe lines can be a failure point when forming. Figures 9-6a and 9-6b show examples of cracks along the scribe lines. Use a superfine point marker or a pencil if the part will be subject to high stress. High strength materials and tight radius bend will fail along these little stress risers.

Forming heavy plate sometimes requires the edges to be softened and potential crack starting flaws eliminated. In Figure 9-7a, the edges were not softened and the start of a crack is plainly visible. In Figure 9-7b, the edges of the plate were sanded to remove the ridges left from flame cutting. In addition, the corners were rounded slightly leading on the plate surfaces. It's just good practice on flame or plasma cut edges to always skim over with the sander or grinder to eliminate potential crack formation points and to remove the hard scale layer that can damage forming tooling.

Flip the layout for cutting on the vertical band saw. Don't forget about the limited throat on a band saw. You can flip your layout to the opposite side to keep cutting with the stock on the outside of the machine.

Figure 9-6a: A crack along the scribe line.

Figure 9-7a: The start of a crack appears.

Figure 9-6b: A crack along another scribe line.

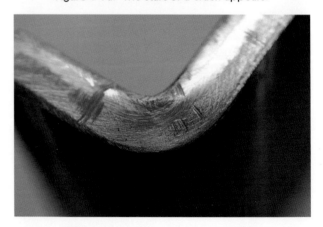

Figure 9-7b: The edges were sanded.

Figure 9-8: A simple centering bar.

Figure 9-10: Placing the center point.

A simple centering bar speeds up the all-too-common task of finding the centerline of strips and bars (Figures 9-8 and 9-9). You can use a transfer punch or a carbide scriber made from a broken 1/4-inch carbide end mill to scribe a line. I made mine with little rollers so you can slide it down the length easily. The rollers are spaced equally distant from the center so the center point is always in the center of whatever width bar is used as long as both rollers are touching.

The center boss is two inches in diameter, which allows you to easily place the center point right where it's needed by burning an inch (Figure 9-10).

Even for a small amount of layout work, a combination square is mandatory. If you have one of those cast aluminum combination squares, do yourself a favor and throw it away right now. Better yet, break in into pieces so nobody pulls it out of the trash can where it belongs!

You can judge peoples' commitment to their profession by the tools they use. If you don't already have one, get yourself a professional forged and hardened combination square set with an additional long blade (Figure 9-11). These tools last so long that your grandkids will be using them long after you're gone. Normally we use the long blades as straight edges or to set distances off an edge. You have to be a little careful and have a fine touch when using a long blade to check squareness. I like several squares when doing repetitive layouts. All I need to remember is small square small dimension, medium square middle dimension, and big square big dimension. You get the idea.

Use a centering head to transfer a line ninety degrees around a corner (Figure 9-12). If you're placing a cross member in relation to your layout mark, be sure to indicate which side of the line the part sits. I mark the correct side with a big X

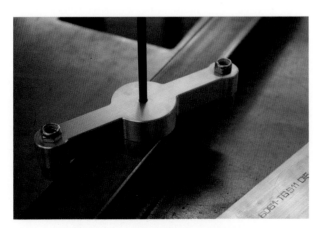

Figure 9-9: Finding the centerline.

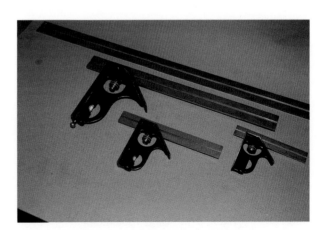

Figure 9-11: Combination square set.

Figure 9-12: Using a centering head.

Figure 9-15: A face extender.

(Figure 9-13). There is not much else worse than having to cut out a fully welded cross-member that's in the wrong place.

Watch out for large tube radii when using your combination square. When you get up to about a quarter-inch tube wall thickness, a standard combination square is not reliable because it's sitting on a curve (Figure 9-14).

Figure 9-15 shows a face extender I made for working with large tubes. It gives me a wider face for those large tube corners (Figures 9-16 and 9-17). It clamps to my standard combination square head with pointed set screws that just catch the lip of the square and push the two faces together. Make it out of something durable so that, when you slide it along the sides of tubes, it doesn't get knarfed up.

Figure 9-13: Marking the correct side.

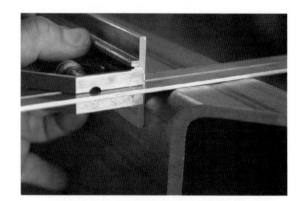

Figure 9-16: Using the face extender.

Figure 9-14: Large tube radii.

Figure 9-17: Working with large tube corners.

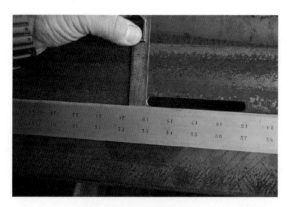

Figure 9-18: Using rules.

Figure 9-21: Working with flexible rules.

Rules are generally superior to tape measures for layout accuracy (Figure 9-18). Quality rulers can be combined with calipers to take large measurements with less uncertainly (Figure 9-19). The overall length of a precision ruler is quite accurate.

Flexible rules come in handy when you need to measure along a curved surface (Figure 9-20) or sneak into a weird spot (Figure 9-21).

PI tapes can be used with excellent results in the fabrication shop (Figure 9-22). One use is to verify parallelism to a very close tolerance. Alignment of parallel rolls or tubes is a good example (Figure 9-23). Another great use is to verify the actual diameter of an object that is out of round. It is much more accurate than taking fifty two-point measurements and averaging them to get

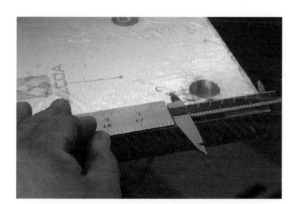

Figure 9-19: Combining rulers and calipers.

Figure 9-22: PI tapes.

Figure 9-20: Measuring along a curved surface.

Figure 9-23: Aligning parallel rolls and tubes.

Figure 9-24: Using the proper level.

Figure 9-26: Comparing two surfaces.

a diameter. This works especially well for rolled sheetmetal tubes and pipes which are less than perfectly round most of the time.

Your layout table should be properly leveled. This allows you to use a precision level to verify geometric relationships that are difficult to check any other way. A level sensitive to at least .005 per foot should be used (Figure 9-24). Accurate electronic levels are available now that have the ability to set their origin on any surface and compare another surface (Figure 9-25). Angular measurements can be some of the most challenging measurements to take. If your table is skee-wacked, then your work may end up that way also.

You can use the origin reset in Figure 9-26 to compare two surfaces that might be very difficult to check with normal angular measuring tools.

A simple auto-level or transit can be used to check geometry, or to level large fabrications (Figure 9-27). These can easily read to .010 over thirty feet. If your datum surface is properly leveled, this is an excellent way to check geometry. This technique is useful for checking features that are not in direct line with one another.

A small bevel is a handy tool for transferring angles to tight spaces (Figure 9-28, third from left). It is a transfer tool. Be careful you don't bump it because there are no marks or divisions on it to verify settings.

Accurate angular measurements are sometimes the most difficult measurements to make. You will need a variety of tools to get the job done, including trammels and protractors (Figure 9-28, first and second from left), and bevels. You will also need a solid understanding of geometry.

Figure 9-25: An accurate electronic level.

Figure 9-27: Checking geometry.

Figure 9-28: Useful measuring tools.

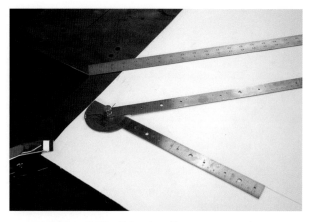

Figure 9-30: Using length measurements.

Large angles are sometimes best constructed geometrically using length measurements (Figure 9-29) instead of using a protractor beyond its basic range (Figure 9-30). A fraction of a degree reading error at a large radius means big trouble. You can resolve approximately a 1/4 degree on a good protractor on a good day with youthful eyeballs. At a 24-inch radius, that 1/4 degree becomes plus or minus .10 inches. Be sure to verify your angle math with a measuring tool that has a scale of some sort on it.

Angular principles, like the easy to remember 3-4-5 right triangle relationship, are an excellent example of a graphical construction. These types of graphical angle constructions are seen in sheet metal pattern work. It's still easier and more accurate for larger work to use the graphic methods for

pattern developments not cut by computer-controlled equipment.

Figure 9-31 shows an example of a typical cutting list. You can see that it has the type and size of each material and any special end conditions. You don't want to be thinking about this stuff when you're at the saw or the shear. I bet you can understand my

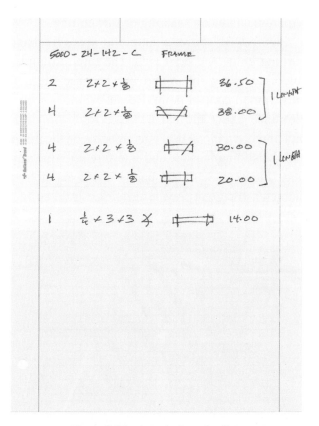

Figure 9-31: A typical cutting list.

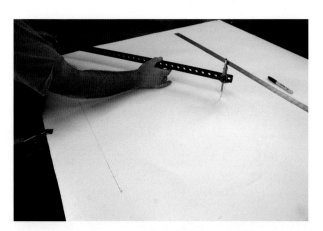

Figure 9-29: Constructing large angles.

Figure 9-32: A freehand burning guide for short cuts.

Figure 9-34: A guide wheel that clamps on the cutting tip.

graphic code for the end conditions of each tube. This thinking should already be done when you step up to the machine. In many of the shops where I worked, there was sometimes a line to use the shear or the saw, so you really needed to get in there and get out as quickly as possible. If you were scratching your head doing any thinking, you got sent to the end of the line. Notice the note for the number of sheets or bars. Be aware of the yield requirements when shearing or saw cutting.

Figure 9-32 shows a simple-to-make freehand burning guide for short cuts. The torch tip slides smoothly along the round bar because there is only a small contact point. Because there is a little clearance under the bar near the torch tip, little bits of molten metal don't get in between the guide and the sliding surface and interfere with the cutting (Figure 9-33). The bar is about an inch in diameter. It's great for those quickie little cutting

jobs and beats the lame, angle iron-type guides hands down.

For longer cuts or where I have to reach across a distance, I made a guide wheel that clamps on the cutting tip (Figures 9-34 and 9-35). The edge of the wheel is a 90-degree knife edge so I can make turns easily. A bronze bearing and a shoulder bolt make up the pivot.

Figures 9-36 and 9-37 show the best circle cutting attachment I have ever used. I almost forgot to mention that I invented it and, no, I don't think it's the best just because I made it. The circle diameter is adjusted by sliding the body along the gas mixing tubes. If more adjustment is needed, you can also adjust the pivot point. With the pivot point shown, you can cut circles as small as 3/4-inch up to a 12-inch diameter circle. Like all circle cutting attachments, it takes a bit of practice to cut a smooth full circle in one pass. It's even harder to

Figure 9-33: Avoiding interference from molten metal.

Figure 9-35: This guide wheel is for longer cuts.

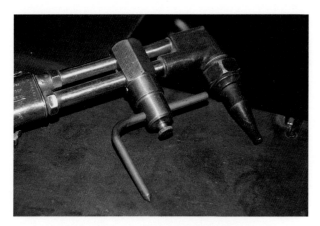

Figure 9-36: A circle cutting attachment.

cut a smooth circle while trying to take a decent picture of it for a book you're writing. . . .

Figures 9-38 shows a great little tool used to stabilize small parts for precision welding. I call it a welding duck. The peg on the top is used to add weight for a little more pressure (Figure 9-39). The flat bar tail keeps the tool from tipping.

Figure 9-39: The peg on top provides more pressure.

The handy little magnetic TIG torch holder in Figure 9-40 prevents you from laying your torch cable over hot material. It will pay for itself in three months easily by saving on broken torch back caps and melted water lines.

Don't worry about the grooves in the grinding wheel used for grinding tungsten electrodes. You

Figure 9-37: The attachment is adjustable.

Figure 9-38: A tool for stabilizing small parts.

Figure 9-40: A magnetic TIG torch holder.

Figure 9-41: Grooves wrapping around the electrode.

Figure 9-43: A block for holding everyday tungstens.

can sharpen tungsten faster when the groove wraps around the diameter of the electrode (Figure 9-41). You should have a dedicated tungsten grinder in the welding shop with a good silicon carbide green wheel on it. If your boss yells at you for putting grooves in the wheel, tell him he's too cheap and should buy a grinder just for the TIG welders.

I never liked the belt sander for dressing tungsten electrodes (Figure 9-42). I came from a shop where we used short back caps, so the electrode started out pretty short to fit in the torch. In most belt sanders, short tungstens are hard to handle to get a close-to-true vertical scratch pattern. The curvature of a normal bench grinder wheel naturally provides hand clearance while grinding.

Sharpen both ends of your tungsten electrodes and have a few close by in the welding area. This

saves dozens of trips to the grinder in a typical week. You can make yourself a cool little block to hold the everyday tungstens (Figure 9-43). Just don't run through the house with this little bed of nails!

I got into the habit of using gas lenses almost exclusively (Figure 9-44). The larger cup size makes a great guide when welding miles of sheetmetal seams, let alone the ability to hang the tungsten out to reach into those tight corners or recesses (Figure 9-45).

With the large diameter cup of the gas lens, you can rest the cup directly on the seam and traverse along without even having to rest your hand on the hot metal or any other support (Figure 9-46).

With a gas lens, you can extend your electrode to reach into those tough corners (Figure 9-47) and still maintain your gas shielding around the tungsten.

Figure 9-42: Belt sander for dressing tungsten electrodes.

Figure 9-44: Gas lenses.

Figure 9-45: Hanging tungsten into tight corners and recesses.

Figure 9-47: Extending the electrode.

Figure 9-46: Resting the cup directly on the seam and traverse.

You try and break it!

At one shop where I worked in San Francisco, we were so busy that it became necessary to hire a few welders to keep up with mounting orders. I was not in charge of hiring so I don't know what kind of advertisement they put out, but from the cross section of humanity that came through the door to apply, it must have been a good one. It must have read something like, "Can you fog a mirror? If so please call 555-555-5555 for a job. Or, check your heartbeat. Got one? Come on down."

My old teacher and foreman Doug used to administer the verbal interview and dole out welding tests to the applicants who were able to fog the test mirror. Like all shop workers, we were very keen and interested in the applicants because they represented potential competition to us all.

"Hey! Look at that one, what hole did he crawl out of? Looks like his hair is dyed—waddaya think?"

The standard test was a couple of simple TIG welds, one fillet weld, and one corner weld in two material thicknesses. By the way, I still use a variation of this test today to weed out the chaff from the grain. Nothing like a fillet weld in thin stainless to separate the meat eaters from their scampering quarry.

Typically the applicant was given several sheared strips of stainless and asked to weld them together as instructed. They were then left alone while they worked on the test, to allow them to get their bearings without the foreman breathing sardine breath down the back of their neck. They submitted the best welding samples when they were done. Some of these so-called welders couldn't get past the tacking of the strips, let alone the test.

Now, between you, me, and the lamp post, this test should not take more than ten minutes if you know what the heck you're doing. It was usually a bad sign if a testee asked for more test strips or a hammer, or if they made more than two trips to the grinder.

Doug normally came back for a quick drive-by pass after fifteen minutes to see how the person was doing. If they needed something or were finished, they weren't standing around with a finger in their back pocket. On the other hand, if they were hopelessly lost, he would escort them out the back door and send them on their way.

Of course, being concerned fellow metalworkers, we might offer what could be construed as help from time to time. An example of the kind of help I mean might have been to drop an empty metal garbage can at the exact moment the poor guy was trying to flip his hood down. Or choose that optimal moment to, say, hammer incessantly on a hollow metal gong.

One day a real interesting character came in for an interview and welding test. I never heard his real name, only the nickname he managed to earn in less than an hour in the shop. This guy looked like he had slept in his car or under a nearby freeway overpass. He had the kind of outdoorsy look to him—and I don't mean camping. To his credit, he was supremely confident and apparently could talk a pretty good line because he was allowed to take the welding test. He must have known a little something about welding or metalwork. I can't believe that the foreman took any pity on the guy and just let him take the test.

Doug got him all set up and even loaned the guy his welding hood. This must have been quite a shock on its own. I picked up his welding hood one time to make a quick tack; it must have had a magnifying lens similar to the Hubble space telescope in it. My little tack looked like the surface of the sun through a pair of binoculars. The tip-off should have been the added weight of a normal welding hood plus the installed telephoto lens system.

After a while, Doug did his drive by to see how this guy was doing on the test. I happen to see it when Doug came back to the welding booth to watch his progress. Doug took one look into the area and shook his head in despair, at the same time suppressing an outright laugh. He yelled at the guy to hold on just a second before he got going. Quickly I ducked around the back of the welding screen to see what was happening.

He had started with the fillet weld. Doug was trying to let the guy down easy and get him out of the shop. He was telling him that he really needed a lot more practice and we really needed an experienced welder. To his credit, the scruffy guy was going toe to toe with the foreman and arguing his case.

When I looked around the corner to see what was up, the guy had the Tungsten electrode hanging out of the torch about six of its seven-inch length. He was actually trying to strike an arc like you would if you were stick welding. The really scary part was that he had struck and scratched so much that the two pieces were actually stuck together.

Doug said in the gentlest of the three tones he spoke in, "Sorry chief, we really need somebody with a little more welding experience. Maybe if you go and practice a few more years, we could give you another test in the future." The guy snapped back, "I already know how to weld. What I need is a job. Go ahead, you try and break it!" With the utmost in patience, Doug said that his welding was not up to company standards and could he please have his hood back.

This guy must have really needed a job badly. I felt sorry for him on the one hand, and some bit of admiration that he had the guts to get in there and actually try to pull off a huge boondoggle.

As Doug shooed him out of the shop, he was still saying,

"You try and break it! Go on, just try!"

So from that point on, if anybody did a slightly-less-than-perfect weld, the comeback was, "Go on, try and break it!"

Figure 9-48 shows a cheap-to-make scriber that will actually cut through the mill scale on hot rolled bar and plate. Most shops have a bucket of wasted reciprocating saw blades available—so who cares if you lose one?

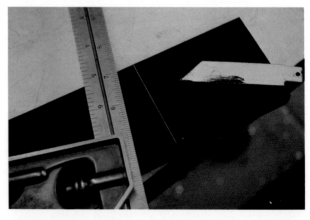

Figure 9-48: A scriber that cuts through the mill scale.

Some of My Favorite Hand Tools

Tools are fun things to collect, but let's face it; there is only a small handful you really use repeatedly on a daily basis. The tools that fall under this heading are almost an extension of your hands and arms. You know just what you can and cannot do with them. I know within a few foot-pounds how much torque I can put on my favorite pry-bar before the tip snaps off. I think about it like this:

A truck with a big NASA insignia pulls up in front of the shop one day and several official-looking guys with dark-rimmed glasses and crew cuts get out. They tell you that the space station has a problem and you are the only one that can fix it. As your head swells upon hearing this, the NASA guys tell you to pack only your most necessary tools for the critical repair mission. You are given a

Figure 9-50: The best rawhide hammer.

strict weight limit for yourself and your tools. Since your belly hangs over your belt a bit, it cuts down on the number of tools you will be able to take. You have to decide in less than three minutes which tools you absolutely must have with you or you will miss the launch window and the space station will slowly spiral into the atmosphere and burn up. This is how I figure out which tools are my favorites.

Garland Rawhide Mallets. These are the best rawhide hammers ever made (Figures 9-49 and 9-50). You can't beat one of these hammers for essentially non-marking hitting power. What gives them the powerful punch is the cast iron head assembly. The faces are available in several material types and are readily replaceable. My two favorites are the smallest #1 and the big monster #5. You will be surprised at what you can move with one of these hammers. Whether you're repairing a global positioning satellite or knocking the dried "residue" off a sewage auger, this is the best soft faced hammer money can buy.

What kind of shark would you be?

When you look back at all the people you have worked with over the years, a few folks stand out. Some you remember because they were very skilled or talented; others you might remember because of some glaring personality trait or defect. There is a special place in my memory for the guys who could never seem to get it right. Everything they touched they somehow managed to foul up or make worse.

Figure 9-49: Garland rawhide mallet.

It was surely not for lack of trying. My old friend Steve was this sort of guy.

At one sheetmetal shop where we worked together, there were two basic departments. One worked primarily with smaller projects in stainless steel and the other group almost exclusively worked in steel. The steel projects were always much larger and included things like large enclosures with hinged doors and covers and lots of grinding.

Typically, each fabricator would shear, form, and do all the weld assembly for a project. The enclosures had a standard sill width that was repeated in almost all the products the company made. The sills were formed as part of the sidewall of the enclosure. It was assembled kind of like a large picture frame with mitered corners.

This is where our buddy Steve first got into trouble that day. He was assembling a large enclosure and had just finished welding the miter joints in all four corners. At some point he realized that each corner had drooped while he was welding it. I won't burden you with the details; let's just say it was a bad thing. This particular enclosure was a little different than the standard type because it was about waist high when laid on its side whereas all the others were nearer four feet.

I never saw him start working on the corners and up to this point was happily minding my own business. I worked back in the corner on the opposite end of the shop. Noise in a working sheetmetal shop is a common thing. You don't really pay much attention to the occasional loud hammering noise until it becomes repeated and rhythmic. It started out as bong, bong, bong, and then a little break. A few seconds later bong, bong, bong, bong. Then a pitch higher, bang, bang, bang, bang. Obviously somebody was having some trouble with something and they were steadily increasing the hammer blows and force with each application.

In most shops, this is the equivalent of screaming out loud, "I just screwed up. Won't you all join me for some humiliation?"

Always eager to oblige, a few folks made their way over to the scene of the crime. Sure enough,

the corners had dropped as they were welded. After trying to grind the exposed weld off, it was pretty apparent that the entire corner would have to come up. A few suggestions were passed to Steve, which he rejected. "This rawhide mallet is working; I just need to get a better swing at it." A few more suggestions were offered when we heard how he intended to remedy the problem.

All the gathered spectators wandered off lest they become somehow associated with this mishap. After a short pause, the hammering continued unabated for the next hour. I guess he was making progress because the pitch of the hammering would change after a while. I assume because he went over to another corner of the enclosure to pummel it into submission.

After an hour of this hammering, everybody was pretty annoyed. I think a few guys had gone over to try to help Steve get out of the hammering thing and onto some other method. About this time, I recall a particularly vicious set of hammering, it sounded like the last corner was not cooperating with Steve.

BANG, BANG, BANG, BANG. Then the last one I heard was BANG, baaaa. The noise level dropped to near dead silence. I really didn't think much of it at the time, but it was all going to make sense in a few minutes.

After a few minutes I saw the foreman come out of the bathroom with a grin on his face and a chuckle in his voice. As he walked past, I asked him what was up. He jerked his thumb back toward the bathroom and said, "Go ask Steve what kind of shark he is."

I buzzed over to the bathroom, a little confused to see what had happened. Steve was peering in the mirror examining his face. As I came up behind him he turned around and looked at me.

Now just to be fair and a little compassionate, I didn't start laughing right away. I waited a second for the scene to sink in. Steve had two rolled-up pieces of toilet paper sticking out of his nostrils. There was some blood soaked into the end closest to his nose. The two white twists of paper looked *like some kind of weird warthog tusks spouting from his nose.*

"What the hell happened to you?" I asked him with the start of a grin on my face. Steve's answer had that kind of sound that your voice gets when you hold your nose closed, "I hit myself with a hammer." I was confused at first—how could you hit yourself in the nose with a hammer? I have managed to smack myself in the fingers and hands a few times, but I have not ever even come close to clocking myself in the face.

Thankfully he volunteered the details of his mishap without me asking. "I was inside that cabinet I've been working on. I was knocking the mitered corners up because they drooped when I welded them." After stating the obvious, he continued with, "I had the big rawhide mallet (#5 Garland) and I was swinging two handed with all I had upward between my legs to get the corners to move at all. I swung and missed the corner completely."

A little snicker squirmed out of my mouth. "You missed the corner and hit yourself in the nose with the mallet?" I asked just to be sure I heard him right. "Yeah, yeah. I know, pretty dumb move," He replied with a note of disgust on top of the already nasally voice.

I had just remembered Doug's question "Hey Steve, what kind of shark are you?"

"A hammerhead—Bite me!"

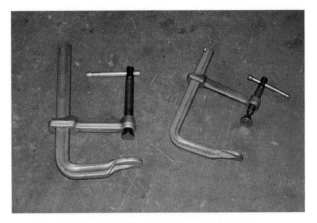

Figure 9-52: Sliding bar clamps.

Bessey sliding bar Clamps. The sliding bar clamps shown in Figure 9-52 came on the scene about 20 years ago. The combination of deep throat depth and excellent clamping power makes these clamps a favorite. You can rapidly position them over a wide range and even use them with one hand in a pinch. Try adjusting a C-clamp with one hand occupied. They are virtually indestructible; their only minor shortcoming is they are a little expensive.

The high-quality clamps in Figure 9-53 come in a wide variety of sizes and abilities. In the welding shop, they are worth their weight in gold.

Merit Grinder with cutting disc. A small, right-angle pneumatic grinder with a good high quality cutting disc on it covers a myriad of odd jobs in the welding shop (Figure 9-54).

Figure 9-51

Figure 9-53: High-quality clamps.

Figure 9-54: Right-angle pneumatic grinder.

Figure 9-56: Steel welding table.

Figure 9-55: Cutting discs.

Pneumatic tools are more compact than equivalent power electric tools. If oiled properly, they will outlive their electron-powered counterparts under heavy duty use. I like the thinner 1/32 thick cutting discs for sheetmetal work and the thicker 1/16th longer lasting discs for weld shop work (Figure 9-55).

Welding Table

The best material for a welding table is cast iron. The weld berries don't stick to them and when you are sliding steel weldments around on them, they move like they are on bearings. The drawbacks are cost and limited configurations. Keep your eyes peeled for an old planer or large milling machine bed. They are tee slotted for clamps and make a great welding table if you can find one. The only

real drawback with cast iron tables is you cannot tack weld temporary fixtures or stops to them. This is a minor inconvenience in my opinion.

Stainless steel on the other hand is just about the worst material you could use for a welding table. Its low thermal conductivity makes the weld spatter, and berries stick without mercy. If you must use stainless steel for specialized work, make it a secondary removable plate that sits on top of your normal welding table.

A steel welding table is the most logical choice for most everybody (Figure 9-56). The table can be readily cut and configured to almost any size and shape. The possibilities are endless.

Install flat ground strips from the machine to the table (Figure 9-57). These are easier to sweep

Figure 9-57: Flat ground strips.

around and cut down on ground cable damage and tripping hazards. If you have expansion grooves in your concrete, these strips can sometimes be inserted edgewise into the expansion to make for a below flush installation.

A Blanchard ground flat surface is best for accurate work. Be sure to stress relieve the plate before grinding to maintain the flatness you grind in. The steel welding table is your basic reference datum. If it's humpy and bumpy, it makes it that much harder to do good work.

Grind both sides. The grinder typically dusts the opposite side anyway to get the plate to sit stable on the machine. As long as you don't specify nutty tolerances and finishes, the price for grinding a large table is not too bad. If one side gets beat up just flip it over and mount the legs on the opposite side.

Check with your local grinder before you order the plate to see how big a plate they can handle if you're thinking about a large table. It's the long corner-to-corner diagonal dimension that matters.

Bolt the legs on so you can re-grind the table surface in the future. Drill the holes all the way through so you can flip the table later on.

Provide overhangs along the edges for clamping (Figure 9-58). Put the legs and any stiffeners far enough in that they clear your deepest clamp's throat.

Figure 9-59: Round holes and square notches for clamping.

Put adjustable feet on the legs. This allows you to push two tables together and get them to match heights exactly.

When you order the plate for your table, ask for a couple of round holes and square notches in the plate (Figure 9-59). These are very handy for clamping out in the middle (Figure 9-60) as well as dealing with the different types of fabrication that come up. There is no magic recipe for the size or locations; just make sure you have one or two.

Put a couple of tapped holes in the plate in a few places. This helps when lifting or handling the table and can be used for add on fixtures or work stops.

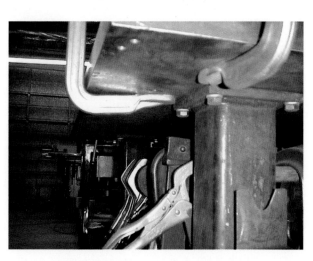

Figure 9-58: Overhangs for clamping.

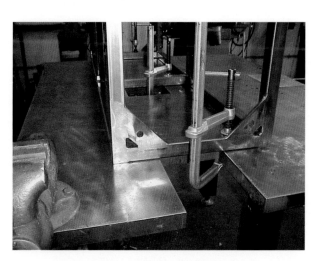

Figure 9-60: Clamping through cutouts.

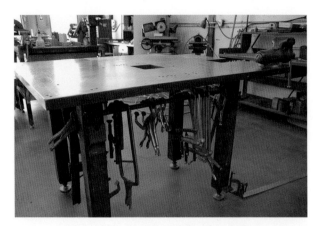

Figure 9-61: A bar for hanging clamps.

Figure 9-63: Using copper jaws.

Put a bar between two of the legs to hang your welding clamps on under the table and out of your way (Figure 9-61). Do this on two sides of the table. When you remove clamps, you will need a place to hang them without walking around to the other side.

One of the best tables I ever worked on was in a shop that used to be an old gas station. The hydraulic car lifts were still in the floor and the main welding table was attached to the top. You could raise and lower the table depending on what you were working on. The anti-rotation rod was removed so you could actually spin the entire table with your project on it. The only bad feature was that, when you had to push or pull on something, the table wanted to rotate. I think there was some way to lock it, but I can't remember how it worked.

Figure 9-62: Using sawhorses.

The heavier the work is, the thicker the table. Ours are 1-1/2 thick for serious stability. Generally for extremely heavy work, the tables are much lower than standard table height, and heavy duty stands or sawhorses are used (Figure 9-62). Be sure to level your stands when you start a big fabrication job. A strip of duct tape on the top of the stand protects your high finish work from scratches.

I always have a vise on my welding table. Make it easily removable if you need the entire table surface. It's worth repeating here to replace the stock jaws with new ones made specifically from copper (Figure 9-63).

Here is a great little story about a welding table I call:

"Put your finger right there!"

At one place where I worked, they had a beautiful cast iron welding table. It was 5-feet wide and 12-feet long. The top surface was deeply ribbed underneath 12-inches deep. It started out life as a large surface plate for assembling diesel engines. The owner bought it at an auction for $100. Two guys could work on it at once with a welding screen in the middle.

At the time I was the new guy, so all the crummy jobs got handed off to see if you cracked. One of these dreaded jobs was building a run of stands. These were angle iron stands, maybe a 3-foot cube in size. They weren't hard to build, but when they built them they made a bunch. I think a normal run was a hundred stands. Needless to say, it was not a real popular job.

I was given the order to make a hundred stands. I worked on them for what seemed like weeks. When I finished almost immediately, I was given an additional run of one hundred stands of a slightly different configuration. At the time I didn't know it, but I was given the additional stands because I had set a new time record building the first lot.

I was pretty bummed at facing another run of these stands. Part way through my esteemed stand, predecessor Steve came up to give me some constructive harassment. First rule of shop survival,

"If they know how to get under your skin, they will—and frequently."

I responded to his ribbing by saying something like, "They gave this job to me because you screwed it up so many times." This went back and forth a couple times until I said "Put your finger right there," pointing to a spot on the welding table. In my other hand I was holding a small ball pein hammer. He said "So what, you're going to hit my finger with your hammer?" I responded, "Yep! I sure am." Well, the dope put his finger right on the spot on this 5-ton table.

This was one of those character-defining moments. If I didn't hit his finger, I would have backed down and he would have won. If I did hit it, he may have responded in the same way and them all hell would have broken loose. I carefully calculated my options and potential weenie factor.

Then I whacked him.

I remember it well. It made a sound kind of like tenderizing a calamari steak. A little satisfying meaty thud. My reputation was signed and sealed at that moment.

After I hit him, he yanked his pulsating finger back and yelped, "I can't believe you hit me!" I responded with a dumbfounded look and a "Umm, what the heck did you think I was going to do?" I really didn't hit him that hard. It was hard enough to make my reputation, but not hard enough to break his finger. I could hear him telling the story

to somebody else a while later, and whoever he told started laughing at him when he got to the part where I whacked him.

This isn't the end of this story.

A few days later he came back after nursing his finger back up to diggity dog strength. I was still building stands on this wonderful welding table. When he walked up, I had been adjusting some stands with a #5 Rawhide mallet. This was one of those big mallets with a cast iron head and 3-inch diameter rawhide face. It weighed about ten pounds.

"This is a Garland number five rawhide mallet, punk. And it will blow your head clean off. You have to ask yourself a question. Do you feel lucky? Well, do you punk?"

He came up and said something like, "So, how are those nice stands coming? I hear they have another batch waiting for you after these." "Steve", I said, "don't you have something better to do than hang around over here getting your finger mashed?" "Yeah, why don't you put your finger on the table and see if I'll hit it," he shot back. "Well, Steve, I'm not that dumb. I'm pretty sure you would hit it after what happened to you. I'll tell you what, why don't you put your head right there." I brushed off and patted a nice cozy spot on the table. "You wouldn't hit me it the head," he said confidently. I couldn't believe my eyes at what happened next.

He actually laid his head on the table.

Fortunately, he was looking away from me or he would have seen the "what should I do now" look on my face. "Okay," I said. "Here it comes!" I picked up the big rawhide mallet and did kind of a mock swing to his head. I made sure I pulled up short. I really didn't want to hit him in the head. So what does the ding dong do but lift his head off the table just as I'm coming to a stop a few inches above him! He thumped his head right into the mallet. DOINK!! This time it sounded like testing a fat ripe watermelon at the supermarket—like a hollow water-filled container.

He squealed in disbelief. "I can't believe you hit me in the head!" I immediately jumped on the opportunity, "Well Steve, when are you going to learn that I mean what I say? You're two for two right now. Care to go for a third?"

He sulked off after this, rubbing his head a bit. It really didn't hit him that hard, but I guess it was hard enough to make him wonder a bit. I am still amazed to this day that he didn't pummel me. I guess there was enough uncertainty and erratic behavior on my part to give him pause.

Later that day a different guy came up to where I was working and gave me a little gloating harassment about building stands for so long. Steve came up shortly after to enjoy my flailing at the hands of one of his remora-like minions. After a while, I ask the guy to put his finger on the table. Steve rushed forward, eyes bulging out of his head to save his comrade and blurted out, "Don't do it! He's wacko! He'll smash your finger for sure!"

After that, I didn't have any trouble at all. By the way, I still have that little ball pein. Almost every time I pick it up, I think about that satisfying noise it made when Steve's finger was compressed between a 5-ton cast iron welding table and its face.

Figure 9-64 shows a low table for heavy steel shop work. The stands are adjustable for precision leveling. This type of table can be easily combined with other tables to make a large flat area. Notice the round bases on the legs. It's easier to tip and roll a heavy leg like this with a round base.

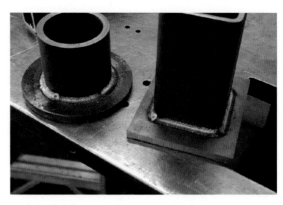

Figure 9-65: Round legs are easier to attach.

Round legs require less welding to attach to a base (Figure 9-65). If you have a large quantity to fabricate, consider using pipe or heavy tube for the columns. The perimeter of a square with the same size is 25% longer than the perimeter of the round. On a big job, that's a lot of extra welding.

Another trick I remembered when I shot the square and round legs photos was if you tack weld from the inside, the shrinking forces tend to pull the plate and the tube together (Figure 9-66). This is an example of using the distorting forces to your advantage. This approach only works in limited cases where you have access to the inside of a tube or the tube has a very large diameter.

Use standard size flat bars for gussets which you can shear directly from strip (Figure 9-67). It's not necessary to have the gusset go all the way into the corner and it takes more prep to fit them up that way. If you must use triangular gussets, then don't cut them to a sharp point. Instead, leave a little

Figure 9-64: A low table for heavy steel work.

Figure 9-66: The advantage of tack welding from the inside.

Figure 9-67: Flat bars used for gussets.

Figure 9-69: An arc starting pad made from copper.

straight at the end to ease welding around the end. They can even double as lifting points if enough space if left in the corner to pass a sling around the gusset (Figure 9-68).

When you have one of those real delicate welding jobs, sometimes it's helpful to use an arc starting pad made from copper (Figure 9-69). This allows the arc to establish itself at a higher current level until you can get your visual bearings without

blowing the whole part away (Figure 9-70). Use copper for your pad; I find it works better if you actually touch the pad with the tungsten before you try to initiate an arc. I have no idea why this works—maybe it breaks the oxide film—but the arc starts more easily when you do this.

Use the ring test to evaluate unknown aluminum filler rods (Figure 9-71). Most shops have two types on hand: alloys 4043 and 5356. When

Figure 9-68: Using triangular gussets.

Figure 9-70: Using the arc starting pad.

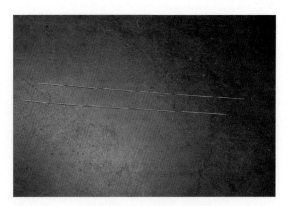

Figure 9-71: The ring test.

Figure 9-72: Imprinting the alloy.

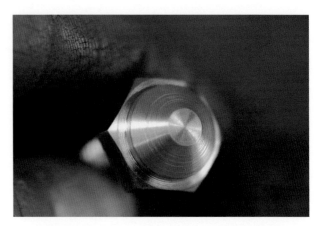

Figure 9-74: A specially-made centering screw.

dropped on the concrete floor, the 5356 rings with a much higher pitch that's easy even for semi-deaf metalworkers to hear. The 4043 sounds quite dull in comparison using this simple test. Some filler rod manufacturers are now imprinting the alloy along the length of the rod (Figure 9-72). This is a great thing if you can read text that small. It only took 40 years to figure it out. . . .

Using your brain and a little welding, you can create some unique composite materials for special applications. One I have used with success is to create a wear surface between two stainless steel parts by buttering one with silicon bronze weld buildup (Figure 9-73). Sometimes called Everdur in the welding industry this composite is then machined to create a sliding interface between the two parts. The example in Figure 9-74 was a special centering screw made from a 316 stainless bolt which was then built up with silicone bronze and machined to size.

Another useful trick involving copper is for heat sinks. If you clamp a copper plate behind a gap you want to fill, it gives you a backing that the weld metal lays on without sticking (Figure 9-75). The copper backing bar absorbs a large amount of heat that would normally go directly into what you are welding, causing additional distortion. This is particularly helpful with thinner materials. The backside of this example is flush with the parent metal and the top surface has an almost flat contour that makes for minimal grinding and no weld prep (Figure 9-76).

Many times a fabricator is called on to make rings of different sizes for various purposes. They might be used for tie down loops or eyebolts (Figures 9-77 and 9-78) or welded to a plate to make a quick pad, or lifting-eye. A simple way to make these is to use heavy stainless spring-type lock washers (Figure 9-79). The twist is taken out of the washer and the joint welded to form a nice round ring of heavy gauge

Figure 9-73: Creating composite materials.

Figure 9-75: Using a copper plate for backing.

Figure 9-76: Minimizing grinding and weld prep.

Figure 9-79: Working with washers.

material (Figure 9-80). It's quick and beats winding heavy rod for one or two rings.

Minimize welding prep by reviewing your design. A few simple changes result in less weld preparation and easier fitup. The angle frames in Figures 9-81 and 9-82 are good examples. They are almost functionally equal. The miter (Figure 9-81) takes longer to

cut and requires weld prep for a decent weld. The butted straight cuts (Figure 9-82) are pretty easy to prepare and produce a frame more quickly.

Structural beams are a real chore to cut notch and fit up in certain ways (Figure 9-83). Here are a couple of examples of an easier way to fit up these types of materials.

Figure 9-77: Using rings for eyebolts.

Figure 9-80: Forming a round ring.

Figure 9-78: Using rings for loops.

Figure 9-81: Comparing angle frames.

Figure 9-82: The advantages of butted straight cuts.

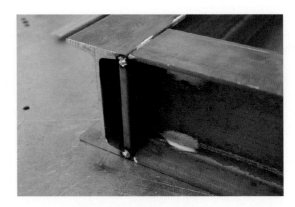

Figure 9-85: Joining the beams.

With the simple addition of flat plates, straight cuts are used to join these beams (Figures 9-84 and 9-85). The plates are easy to cut and add strength to the connection by increasing the amount of weld in the joint compared to the notched fitup shown in Figure 9-83.

Try to think in smaller sub-assemblies. They are easier to handle for welding and have a smaller distortion control problem. The leg sub assem-

blies seen in Figure 9-86 will be completed and straightened before they are added to the larger assembly.

Deep butt weld tube fills on heavy wall material can be a real welding chore. To get a nice flush weld that grinds smooth takes a large amount of welding and filler material (Figure 9-87). An alternate that looks great and saves on welding and the subsequent grinding and straightening is

Figure 9-83: Structural beams.

Figure 9-86: Leg sub-assemblies.

Figure 9-84: Using flat plates.

Figure 9-87: Getting a flush weld.

Figure 9-88: Changing tube size and weld.

Figure 9-90: Using long bolts and studs.

to use a tube one size smaller and fillet weld (Figure 9-88).

Sometimes a fabricator is required to put threaded bosses in a tube wall. Figure 9-89 shows a simple way to handle that without making any special fittings. The hole size is the dimension across the points of a standard or heavy nut. You can hold the nut for tacking using a long bolt. This beats drilling and tapping round bars in the lathe to make custom weld nuts. Be sure to use a long bolt or stud when you install these in a frame (Figure 9-90). With the long stud or bolt, it is easier to get the thread axis square with the world. You can make your own weld nuts obviously, but these are readily available and dirt cheap compared to the home-made models. How many threads do you need anyway?

If the hole is sized correctly, then installation is a snap. Remember: it's easier to sand a little off the corners of the nut than to produce a hole that fits the nut perfectly. For sealed threads, we use heavy nuts

and weld a sheetmetal punch slug over the end of the nut (Figure 9-91). Another use of this technique is to use your entire tubing frame as an air storage tank. I built a frame one time where all the tubes were interconnected for the purpose of an air storage tank. The frame had a lot of pneumatic equipment on it, so pipe thread bosses were installed wherever air was needed on the machine. It was quite a large frame and had a considerable air volume inside the tubing.

Another way to handle tube-to-corner fills is to inset the member away from the radius (Figure 9-92). This leaves us with a much smaller, less distorting weld to deal with.

A simple change makes an annoying welding job (Figure 9-93) a piece of cake (Figure 9-94). Joint access is something that is difficult to predict for designers, but even harder to execute by welders.

Long wire and electrode stick outs are needed to reach into this annoying-to-fill joint (Figure 9-95). The tube wall takes the brunt of the weld heat and

Figure 9-89: Putting threaded bosses in a tube wall.

Figure 9-91: Sealed threads.

Figure 9-92: Working with tube-to-corner fills.

Figure 9-95: Using long wire and electrode stick outs.

penetration; on thinner tubes, you risk burn through on the tube ID. It's not too bad with MIG and stick welding, but is particularly important with TIG welding thin wall tubing in stainless steel where you don't have the reach.

A small change to the joint design makes for fast efficient welding. This joint (Figure 9-96) is

about as easy as it gets in the welding shop. A little thought goes a long way.

Make your own custom tubing. The more professional looking method is to use two formed channels (Figure 9-97). When forming the channels, leave the bend a few degrees open to allow for the weld contraction (Figure 9-98). It's easier to

Figure 9-93: Changing an annoying weld job...

Figure 9-96: Adjusting the joint design.

Figure 9-94: ...to an easy weld job.

Figure 9-97: Two formed channels.

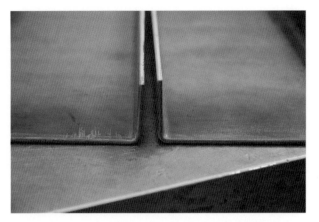

Figure 9-98: Leave the bend open a few degrees.

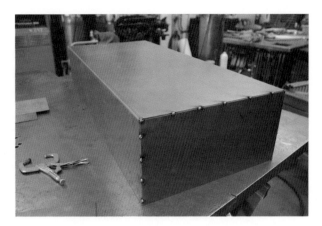

Figure 9-101: Form long, weld short.

flatten a protruding weld seam than to pull up a sunken one. This is sometimes the only way tubing can be obtained in special materials or odd sizes. The trick works well for all those small quantity, short lead time jobs.

Be sure to check squareness when you are tacking the channels together (Figure 9-99). With the

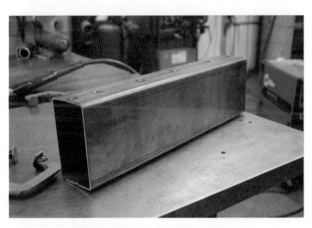

Figure 9-99: Check squareness.

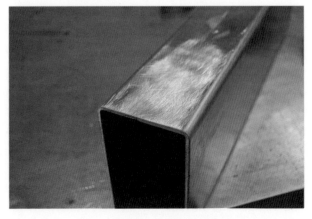

Figure 9-100: Custom tubing.

welds off the corners, the custom tubing has that professional high production look just like the commercial material (Figure 9-100). You can also make tubing from two L-shaped formed parts, but the results never quite look like real tubing.

Form to eliminate or at least minimize welding. If you have a choice, form the longer runs instead of welding them, as shown in Figure 9-101. Remember the rule, "Form long, weld short."

Brake Bumping

You can bump roll cylinders and other radii easily in the press brake. Thick-walled tubes and rings can be formed by this simple-to-learn technique. Many folks have seen large light poles out in public that are formed this way. They show the characteristic cogging or faceting that this method sometimes produces. Fundamentally the method is no different than rolling in a set of slip rolls. With slip rolls, the material is formed in a continuous fashion. Every tiny bit of material is bent by the rolls to form the smooth curvature. When we bump roll in the press brake, we abbreviate the process by only forming incrementally to get an approximation of the full curve. Depending on how many times you care to bump, the end results are indistinguishable from a rolled form. The advantages are that press brakes generally have a capacity for heavier or longer work than rolls and have no real lower limit on diameter. The disadvantages are seen when bumping large one-piece diameters because of the opening of the press

Figure 9-102: Laying out parallel lines.

Figure 9-104: Brake bumping.

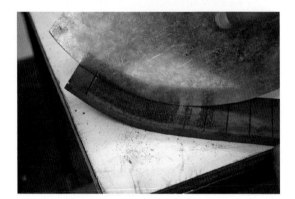

Figure 9-103: Checking progress.

Figure 9-105: Forming perpendicular strips.

brake. The one exception is a press brake equipped with a horn extension.

I start out by laying out a series of parallel lines one inch apart (Figure 9-102). These are the general guide lines that will be used to reference your progress. You will also need a sheetmetal sweep or template to check your progress as you go (Figure 9-103). I always start out with three hits per inch with a light bend. You can calculate the approximate angle of each bend if you feel like it, but I usually just use a test piece to get the radius close. Keep in mind that it's easy to add more curvature, but a real pain to take some out. Therefore, always come up a little short with three hits per inch. To correct the final curvature, you can bump in between your previous hits to add more curvature instead of increasing the amount of each individual bend. This produces more predictable results than increasing the bend angle by deeper punch depth.

This method of brake bumping is also used to pre-form the ends of plates and sheets that are going to be rolled in conventional rolls (Figure 9-104). With many power rolls, there will be a large flat spot at the ends of the plate if they are not pre-formed.

The strip on the left side of Figure 9-105 is used as a rough guide to make sure the strip is formed perpendicular to the edge. Typically rings are formed in two halves when they are a large diameter. Because of the way the material is supported in the press brake, we must add some on each end so the end can sit on the die for the forming process. This excess will be removed later when the part is formed to the correct radius. For large or heavy parts, you can suspend the part with a sling from the crane using a clamp or a choker hitch.

The flanged, one-piece cylinder in Figure 9-106 was formed with the press brake bumping method. The flange was formed on the flat pattern in the

Figure 9-106: A flanged, one-piece cylinder.

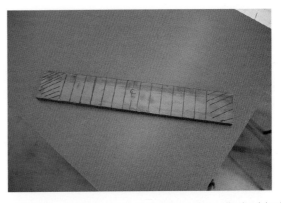

Figure 9-108: Leaving extra material on the cylinder blank.

Figure 9-109: Preparing the cylinder material.

press brake and hung over the end of the die as the cylindrical portion was formed. You can see the faint lines from the bumping on the ID (Figure 9-107). This kind of cylinder can be formed in a set of rolls, but only if they have a deep relief groove in the top and bottom rolls, which most rolls don't have.

Extra material is left on the ends of the cylinder blank in Figure 9-108 so the material can sit on the

wide lower die opening properly for bumping (Figure 9-109). The forming sequence is to form in from each end first (Figure 9-110), and then form the center section last (Figure 9-111). It is much easier to position the part for the multiple strikes used in cylinder bumping if you use this sequence.

After forming, the cylinder is slipped off the special shop-made deep punch. I mark the trim

Figure 9-107: Faint lines from bumping.

Figure 9-110: Form in from each end first.

Figure 9-111: Slip the cylinder off the punch.

Figure 9-114: The type is tacked together.

points with a center punch before I form the blank. After it is partially formed, it's easy to see where the cut will be made. The extra material that was added for spanning the lower die is now trimmed off and the cylinder closed (Figure 9-112).

After the last couple of hits are made to round up the tube, it is slipped off the upper punch and tacked together (Figures 9-113, 9-114).

Material is .50 thich × 4 wide steel. You can fabricate these custom-sized short tubes that are hard to even find commercially, let alone purchase in short lengths. Figure 9-115 shows a special sleeve made to fit an odd-size tube.

Extra-wide lower dies for bumping can be quickly made in the fabrication shop for a rush job (Figure 9-116). Figure 9-117 shows a custom lower die made for bump forming heavy plate.

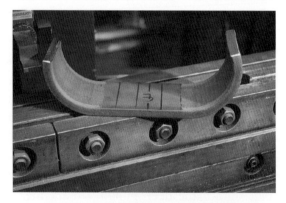

Figure 9-112: Extra material is trimmed.

Figure 9-115: A special sleeve for an odd-size tube.

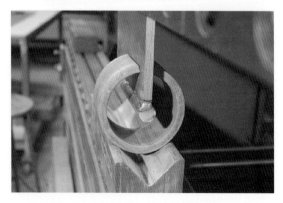

Figure 9-113: The cylinder is closed.

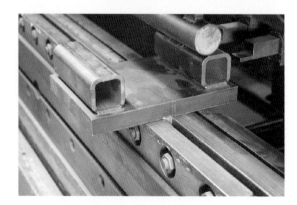

Figure 9-116: Extra-wide lower dies.

Figure 9-117: A custom lower die.

Figure 9-120: Hide the punch mark.

When working with heavier material in the pressbrake, it's sometimes easier to use punch marks to locate the bend lines than program or set the back gage (Figure 9-118). Most back gages in plate shops are beaten to death from heavy plates banging into them. Center punch the ends of the bend lines (Figure 9-119) and just hide the punch mark under the upper punch (Figures 9-120 and 9-121) to locate the bend centerline. Marker

smears and mill scale flakes off during forming, taking any scribe lines with it. Center punch marks will still be there.

Just barely hide the diameter of the center punch mark (Figure 9-121). This puts the cener pretty darn close to the punch centerline.

Here are three methods of capping tube ends in order of speed and preference from worst to best (Figures 9-122, 9-123, and 9-124). The cap fitted

Figure 9-118: Use punch marks to locate the bend lines.

Figure 9-121: Hide the diameter of the center punch mark.

Figure 9-119: Center punch the ends of the bench lines.

Figure 9-122: Capping tube ends.

Figure 9-123: Fitting the end cap outside.

Figure 9-125: Making a frame or window cutout.

for a corner weld has the advantage that it can be finished by grinding to make an invisible connection that looks like the stock tubing (Figure 9-123).

Sometimes it is easier for fitup if the end cap is designed to fit on the outside (Figure 9-123). Fitting the end cap to the inside requires more time and higher precision in each of the two parts, not to mention actually positioning it for welding. The corner-to-corner method requires weld finishing if a blended look is desired.

The method in Figure 9-124 gives a larger footprint, easy welding, and low precision of the mating parts.

Sometimes it's more efficient, depending on the equipment you have available, to make a picture frame or large window cutout by connecting flat strips to make the outline (Figures 9-125 and 9-126). Cutting a large center out of a flat sheet is more wasteful of material than shearing strips and

can have stresses left in depending how the cutting was done. It's hard to beat the clean, neat look of sheared edges on the inside with any manual cutting process.

If you make the frame a little large, you can actually shear it to final accurate size after welding and grinding (Figure 9-127). With good sound

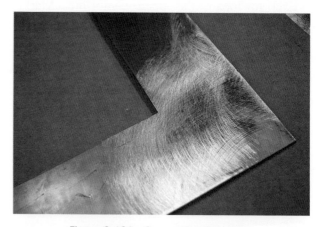

Figure 9-126: Connecting flat strips.

Figure 9-124: The fastest capping method.

Figure 9-127: Shearing a frame if needed.

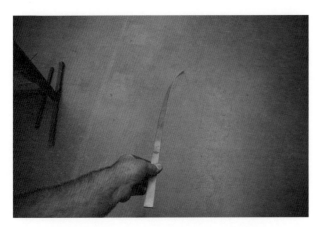

Figure 9-128: Strips get twisted sometimes.

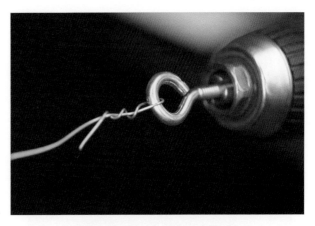

Figure 9-130: Straightening wire.

welding, forming this frame into a door or cover is a simple matter.

Many times when you shear a narrow strip for a project, the shear puts a long sweeping twist in the material (Figure 9-128). The best way to remove the twist is to hold the strip by the extreme end with minimum material clamped in a vise (Figure 9-129).

Figure 9-129: Removing the twist.

By twisting it as a whole with a wrench from the extreme opposite end, you get a perfect straight strip. The shear twists the entire strip when it's cut, so the correction must be the precise opposite of the shearing process. If you clamp short sections and work your way down the strip, it will take longer and the results will be pretty dodgy.

A similar trick works for straightening wire. Clamp one end of a long wire you want to straighten in the vise. Clamp the opposite end in a drill motor or loop it around a small eyebolt held in the drill (Figure 9-130). Pull a little tension on the wire and spin the drill (Figure 9-131). What happens is the wire yields as it is spun and takes on the new shape, which, if you did it right, is nice and straight. A common use would be straightening small diameter MIG wire for use as TIG filler rod.

After you have squared and braced a simple frame (Figure 9-132), be sure to use the proper weld sequence to maintain your squareness. Always make the least distorting welds first and check for movement as you go, not after you have welded it up. Braces can be tack welded or clamped (Figure 9-133).

Always end on a tack weld instead of starting on one (Figures 9-134, 9-135, 9-136). If you start on a tack weld, it softens it and allows movement before the joint has developed some strength of its own from your weld.

In general, leave all fillet welding until last. Fillets distort terribly, but most people like to do them first because they are easy and look good.

Figure 9-131: Spinning the drill.

Figure 9-133: Clamping braces.

Figure 9-134: Ending on a tack weld.

Never weld anything in an unrestrained condition unless you want it to move. The lowly clamp has the ability to reduce your welding distortion by a huge amount.

An easy way to think about it is if you are welding two pieces of flat bar together end-to-end with a gap between them, as shown in Figures 9-137 and 9-138. In the first picture, the two bars are clamped lightly with vise grips so I could measure the overall

length without the bars shifting. In the second picture, I have clamped a second set of bars with a pair of heavy duty, professional welding clamps. Machine settings are identical for both welds.

If you can imagine the weld metal as a tiny turnbuckle that pulls the ends of the flat bars closer together as the weld metal cools, you are halfway there to understanding what takes place (Figure 9-139). Now visualize adding a clamp on

Figure 9-132: A squared and braced frame.

Figure 9-135: Avoid movement.

Figure 9-136: Give the joint time to strengthen.

Figure 9-137: Welding together two pieces of flat bar.

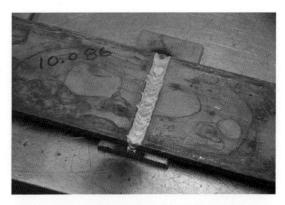

Figure 9-139: Pulling the ends closer.

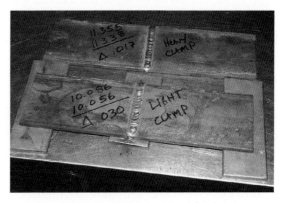

Figure 9-140: Reducing longitudinal shrinkage.

each bar that will not allow the bars to move closer together. The result is that all the weld shrinkage takes place within the weld metal instead of using its shrinking energy to pull the bars closer together. From this simple test, you can see the longitudinal shrinkage was reduced by half just by changing the type of clamps that were used (Figure 9-140). If you were really paying attention, you would have noted that even under

heavy clamping it still moved. Any good welder has learned how to deal with and predict weld distortion. If you remember the simple credo, "Everything moves, always," you're halfway there.

Check your parallel dimensions first when squaring anything (Figure 9-141). You cannot be square if you're missing parallelism first. By comparing the corner-to-corner dimensions, you are using a sensitive geometric proof to verify a square

Figure 9-138: The gap between two pieces.

Figure 9-141: Checking parallel dimensions.

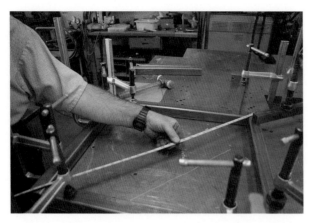

Figure 9-142: Comparing corner-to-corner dimensions.

Figure 9-144: Attaching flat bars.

(Figure 9-142). A combination square or carpenter's square is a more difficult tool to use than the cross corner method when moving at speed for squaring jobs.

If you have a fabrication that needs several accurate surfaces, it's easier to weld a pad for the precision component to sit on (Figure 9-143). This saves having to grind or surface the entire face and the inherent problems with that method. Don't make the mistake and skimp on the machining or grinding allowance of your component pad. Your machine shop or surface grinder might complain a little if they have to remove a little extra material, but it pales in comparison to the whining you will hear from the machine shop if the surface doesn't clean up completely.

If you need a continuous precision surface on an otherwise lightweight fabrication, attach flat bars to provide material to precision surface for flatness or

other geometry (Figures 9-144 and 9-145). Generally, tube walls should not be surfaced directly to create a precision surface because of excessive wall thinning. These flat bars will be surface ground after the frame has been stress relieved to provide a super-accurate mounting surface for some fussy components (Figure 9-145).

Make yourself a few sizes of angle iron V-blocks (Figures 9-146 and 9-147). These are easy to make and handy around the welding shop when working on piping or handrail projects Make a few sizes in pairs for all those round fabrication projects in the shop.

For long V-blocks, you can use old sections of press brake dies. The extra length makes a big difference when you have to connect two longer rounds together (Figure 9-148). Don't use nice new precision dies unless you want to create an enemy in the forming department. Brake dies are precision tooling and should be treated as such.

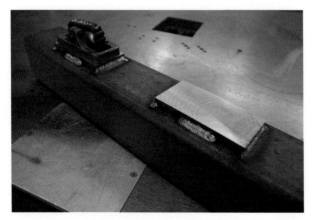

Figure 9-143: Welding a pad.

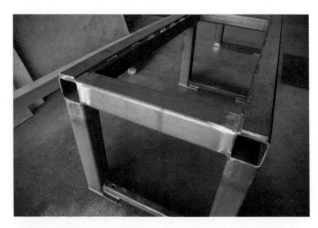

Figure 9-145: Providing an accurate mounting surface.

Figure 9-146: Iron V-blocks.

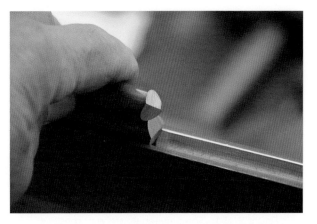

Figure 9-149: An alternate way for prepping the ends of rounds.

Figure 9-149 shows an alternate way to prep the ends of rounds to weld them together. This is the preferred method to avoid lots of stops and starts in your welding. By prepping the round with a standard double bevel, you can do most of the fill without rotating the part every few degrees. For welding processes that produce slag, fewer starts

and stops are important. In big reinforcing rod joints, the weld preps are done this way.

Clamp your ground clamp directly to delicate work (Figure 9-150). Some materials are very sensitive to arc erosion through the ground—aluminum and magnesium are two examples. These marks look really bad and are hard to remove because there is usually metal transfer from the table to the part (Figure 9-151).

Figure 9-152 shows the start of an auger layout. The long centerline is marked on the tube by laying the tube in a length of angle and scribing or marking along the edge. The pitch is marked off along this centerline to guide the tacking and fitup of the flighting sections (Figure 9-153). On larger screws and augers, a come-along or crane can be used to help stretch the flighting along the center tube.

Figure 9-147: These V-blocks are easy to make.

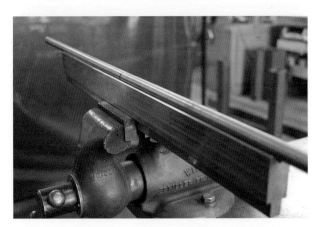

Figure 9-148: Connecting longer rounds.

Figure 9-150: Clamping a ground clamp.

Figure 9-151: Some materials are very sensitive.

Figure 9-154: Stainless filler rod.

You can use stainless filler rod to tack up a steel job, if the job will allow it (Figure 9-154). The stainless tack is much more ductile than a steel filler tack. The parts can be easily bopped around with a hammer to allow precision alignment and fitup (Figure 9-155). The advantage is that stainless tacks are the more ductile material

and don't crack like steel filler when you make fitup adjustments.

I'm sure every welder has fought a porous casting at one point or another (Figure 9-156). Sometimes nothing you do will clean out the embedded contamination and pinholes to allow a decent weld. One trick I have used with success is

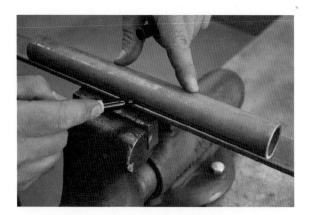

Figure 9-152: The start of an auger layout.

Figure 9-155: Precision alignment and fitup.

Figure 9-153: Guiding the tacking and fitup.

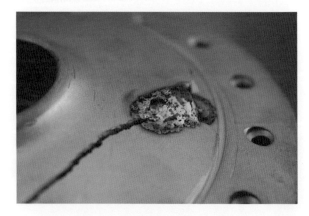

Figure 9-156: A porous casting.

Figure 9-157: Thinking like a dentist!

Figure 9-158: Judicious pre-heating.

to use a drill to actually drill into the offending pockets, kind of like a dentist drilling out a cavity in your tooth (Figure 9-157). The drill removes the buried pockets of crud and gives you a fresh shot at making a decent weld. Another trick along these lines, if you have the facilities, is to cook the casting at a high temperature to leech out the dirt and crud. The temperature will vary depending on the material you are working. As a rule of thumb, only cook the part at a temperature at which you would do a preheat, or maximum interpass welding temperature.

If you have some heavy material to weld, you can reduce your amperage requirements or increase the maximum thickness you can weld with some judicious pre-heating (Figure 9-158). I like to preheat anything over 3/8 thick when we are running smaller wire diameters in the MIG welder. It helps with welding speed, fusion, and re-starts. My temperature gauge is when I see a little smoke from the surface scale or a temperature that will cause leather to smoke a little when touched to the metal.

In this chapter, we looked at many different tricks and techniques used in the welding shop to handle those tricky jobs and work more efficiently. In the next chapter, the discussion will move into methods used to control welding work to a much higher degree.

The Lost Art of Flame Straightening

10

Flame straightening is one of those old-timer skills everybody has heard of but few people have had the opportunity to watch someone do. The ability to straighten bent plates and shafts is almost magical in its simplicity (Figure 10-1). It's definitely a valuable skill to learn if you intend to spend any time around a welding shop, unless you relish the though of beating a stubborn shaft or plate into submission with a mallet. This easy-to-learn technique has been used to straighten everything from the hulls of ships to samurai swords.

Welders and metal fabricators are continually confronted with the negative aspects of joining metals by welding, some welders more than others. Whenever metal is heated it expands and not always in a favorable way. Most often it has a negative effect on the end product. We have all seen the results manifested as warpage, distortion, and stress in our own work. If you have spent any time in the weld shop, you have seen some of these kinds of distortion like welds pulling, flame cut strips and shapes twisting and bowing, oil canned plates, and covers. The list is as long as there are things that are heated during manufacture.

Every welded structure will try to seek the position of least stored stress. A simple example that best illustrates this is when you crumple up a sheet of paper tightly and then drop it on the desk. The paper expands slightly and finally settles in the

Figure 10-1: Straightening bent plates and shafts.

233

position of least internal stress. As we weld on a metal structure, this is happening continuously as we work. The weldment is undergoing a constant re-arrangement to find this state of least stress. Welding sequence, clamping, and fixturing can help, but can never eliminate the effects of expanding hot metal. When we stress relieve a weldment or part, we are allowing whatever remaining locked in stress to dissipate by heating or, in some cases, vibration.

Flame straightening uses the same principles that cause the warpage and distortion to make positive corrections or, in some cases, enhance the effects. The best way to imagine how this process works is something like reverse welding. Figure 10-2 shows an example of using the effects of controlled distortion to induce a large camber. These kinds of results can be accomplished with the humble cutting torch and the knowledge of where to apply the technique. In this example, if I had a set of rolls large enough I would have run the tubes through. These kinds of results can be achieved in Joe McGee's one light bulb garage or standing on scaffolding five hundred feet off the water using flame straightening.

Unlike pressure methods like hydraulic presses and screw jacks, flame straightening is the laser-guided smart bomb that attacks the exact offending areas. Pressure methods distribute their gross application of forces throughout the structure and sometimes cause more harm than good. Hydraulic

Figure 10-2: Inducing a large camber.

rams further complicate the correcting measures because they lack a sense of feel connected with the yield point of a particular material. Unless you have a good pressure gauge to monitor your force application, the yield point can be a dangerous, easily-exceeded target. Pressure methods require the parts to be supported so the force can be applied. It's kind of like pulling yourself up by the bootstraps. I have seen good work ruined by the sloppy application of force methods. I have also seen a guy knock out his two front teeth using a hydraulic ram to straighten a frame, but that's another story all on its own. In many cases, the materials physically cannot bear the loads needed to restore them.

Force methods typically require increased operator handling and labor. With flame straightening, plates can be straightened installed as the hull of a ship or where they lay on the shop floor. Shafts and tubes can be corrected while they're still in the lathe or connected to machinery. The portability of this method alone is worth the time spent to learn it. Flame straightening can be used where pressure methods would be impossible or extremely awkward. This is not to say that pressure methods don't work. As any experienced metal fabricator can tell you, sometimes you need all the tricks in your bag to get the job done. All I am doing is giving you something else to put in your tool bag.

So what is flame straightening exactly? I like to define it as,

> *"The specific application of controlled cycles of heating and cooling that are used to correct, enhance, or minimize distortion in metals."*

Flame straightening causes a contraction or localized shrinking in metals that are used by the applicator (you) to affect changes in geometry. The technique is applied with a common oxy-acetylene torch in a specific way that induces the maximum amount of contraction. Now before you run out to the shop and fire up your torch, you should read

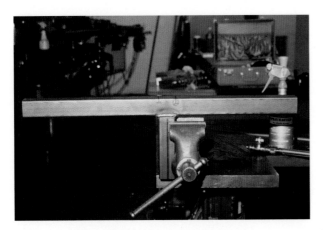

Figure 10-3: Bowing the long tube.

the rest of the chapter. There are a few important points you may want to know about.

Flame straightening works best when the distortion is caused by welding or is heat related. It can be used on any materials that have no restrictions on the application of heat from an oxy-acetylene torch. It works best with materials that do not have high thermal conductivity. Steel and stainless steel are the primary candidates. Aluminum and copper are less than enthusiastic about this technique. You can see in Figures 10-3 and 10-4 the two fillet welds joining the two tubes have bowed the long tube. This would be a typical weld distortion problem encountered in the shop. Some of this bowing could be eliminated by good fixturing and clamping, but for the example I wanted a maximum distortion.

Figure 10-4: A typical weld distortion.

Limitations

Flame straightening shouldn't be used on materials that would be damaged by the application of heat from an oxyacetylene torch. Wood is not a good candidate. A few examples of materials you should avoid are metals that react with oxygen when heated (titanium) and heat-treated parts that have been tempered to temperatures less than 1300°F. Certain kinds of alloy steels, tool steels, and some steel forgings can be damaged by the kind of rapid heating used in flame straightening. In other words, be sure you know what you are working with before you screw something up. Don't say I didn't warn you! This is where welding engineers earn their keep.

How Flame Straightening Works

Metals, like all materials, expand when heated. In the case of steel, it expands .000006" per inch, per degree of temperature change. By using this technique of very localized, intense heating, we can cause what we are looking for—a small cross section of material to expand considerably and form an actual material upset, and then the subsequent mechanical contraction of the material as it cools.

We are trying to induce a large temperature difference between the area we are heating and the surrounding cooler metal. The contraction or shrinking effect we desire is created by the temperature difference or gradient. We are, in effect, creating the same conditions we see when welding: a very hot, intense, local heating surrounded by cooler base material.

As this local area is heated, it is trying to expand rapidly. It encounters resistance from the cooler surrounding material, which creates pressure. This pressure flows in the path of least resistance, which in most cases is upward along the area softened by heating. This upward swelling forms the upset. And as it cools, it contracts and acts like a tiny heavy duty turnbuckle pulling the surrounding material with it. This upset can be seen on thinner materials as bulge or slight swelling. In

Figure 10-5: Bulges left from contraction.

Figure 10-7: Help from a strong-back and clamps.

Figure 10-5, I have intentionally highlighted the slight bulges left from the contraction by dusting them with a sander. Notice that the corrective action was done directly opposite the fillet welds on the tee which caused the distortion.

Heat Input

One of the basic secrets to using this technique is the heat input. The trick is to heat the material as quickly as possible in as small an area as possible (Figure 10-6). If we used a normal rosebud-type heating tip, we cannot confine the heat to a small weld-sized zone. The best tips to use for most normal flame straightening are cutting tips. Only the preheat flame is used to form the upset. These tips are designed to heat a small area quickly in preparation for flame cutting so they work well for the rapid heat input into a small area we are looking for. Rosebud

tips are for burning weeds in back of the shop or large area heating, not tight localized spot heating.

For this example, I had some mechanical assistance from an aluminum strong-back and a couple of clamps (Figures 10-7 and 10-8). This enhanced the effects of the heat shrinks so the work was completed faster.

Figure 10-9 shows the sample after correction. The material is $2 \times 2 \times 1/8$ wall steel tubing. I am pointing to the lack of a gap between the straight-edge and the tube.

Always use the correct fuel and oxygen pressure and flow for the size tip you are using. There is some confusion as to how far to open the fuel valve when setting your flame for a particular size tip. Open the fuel valve and light the torch in the normal manner with the correct regulator pressure settings for the tip size you are using. After you have it lit, open the fuel valve until there is a separation between the base of

Figure 10-6: The heat input is key.

Figure 10-8: Enhancing the effects of the heat shrinks.

Figure 10-9: The sample after correction.

Figure 10-11: Adjusting the fuel valve.

the pure fuel flame and the tip (Figure 10-10). Slowly close the fuel valve just until the base of the flame re-connects with the torch tip (Figure 10-11). This is the correct amount of fuel for the tip. More than this is too much fuel flow and the flame will be hard and forceful. Less than this and you are not heating as efficiently as you could be.

We always apply the corrective heat to the high spot we are trying to remove. If it is weld distortion, we work on the opposite side from the offending weld. The only exception to this is if we were trying to induce curvature or deliberate distortion such as a camber.

Mapping

The first step in any correction job is mapping the high points with a straightedge (Figure 10-12).

For the demonstration, we will map and correct the distortion in a typical mill supplied plate.

You will need a few tools to help you map the distortion. Several straightedges of different lengths and markers that won't wash off with water are used to plot the typical wandering distortion found in plates. I like the faithful old Sharpie. It leaves a hazy reminder of where you have applied shrinks even after the intense heating.

The first step in plotting the distortion is to do a quick check with the straightedge along both axes of the plate to find the worst side, which is the side with the largest number and most severe high spots. For correcting plates, I like to have the plate on horses or blocks on the table. Support the plate on three points so it doesn't rock. The best is outside on the forks of a forklift with a water hose nearby. The older I get, the less I like to bend

Figure 10-10: Opening the fuel valve.

Figure 10-12: Mapping the high points with a straightedge.

Figure 10-13: Plotting in the direction of worst distortion.

Figure 10-14: Comparing the straightedge to the plate.

down. There is a sweet spot where you don't have to bend too far to sight the gap between the straight edge and the plate is not too high to reach across. The older you get, the narrower this range is.

To locate the high spots, we will use our longest straightedge first. Hopefully it spans the entire width of the plate. If you don't have a straightedge long enough, a decent piece of flatbar will also work. Before you begin, decide how flat you are going to make the plate. It's pretty easy to change your standards as you work and end up putting more work in than necessary.

Start your plotting in the direction of worst distortion. Move your straightedge perpendicular to the edge (Figure 10-13). If the plate is round or odd shaped, pick a direction and stick with it for the entire straightening job. As you compare the straightedge to the plate, place your eye at the same level as the gap between the plate and the straightedge (Figure 10-14). This makes it easier to see the gap.

Move the straightedge along the plate. As you find the high spots, mark them with your marker. Mark in approximately two-inch increments as you move along the plate. When you find a high spot, rock the straightedge back and forth to locate the apex. The rocking see-saw motion will show you exactly where the apex is. Mark these apexes as you move along the plate. If there are two high spots or the high spot is broad, mark the points where the change is greatest.

It is very common to have distortion that requires you to flip the plate from side to side. Do as much correction on one side as possible before you flip the plate. It's much too awkward to try to apply your correction in the overhead position. Unless you have something to prove about your abilities as a contortionist, I would just flip the plate. If the plate weighs 25 tons, then maybe it makes more sense to do it from the underside. I got to pick this example plate and, in case you hadn't noticed, it's pretty small and easy to flip for us old guys.

As you make your marks along the plate, use line weight to remind you of the approximate amount of distortion. A light dash indicates a minor variation and a heavy dash indicates a large distortion requiring a full correction. Many times when mapping the distortion, you will find that it fades out and ends abruptly. Mark the end point to remind you to stop your straightening at that point (Figure 10-15). Typically there will be several wandering dashed lines, some ending and some running the full length or width of the plate.

So now we have marked out the variations in the plate we want to correct. We are now ready to apply our correction.

Applying the Correction

The flame straightening correction is called a "heat shrink." It is applied with a common oxy-acetylene torch with a standard cutting type tip.

Figure 10-15: Marking the end point.

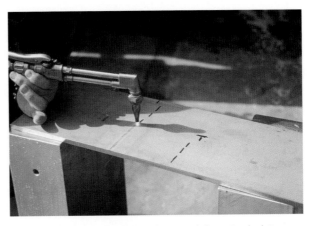

Figure 10-17: Straightening a stainless steel plate.

Tip selection is determined by the material type and thickness. Use a tip the same size you would for cutting when shrinking stainless steel and one size larger than you would for normal cutting for steel. Oxygen Acetylene is the preferred fuel combination. Acetylene oxygen has a higher Btu/min input than other common cutting fuel combinations. Remember we are trying to heat a narrow zone very quickly.

When we make our corrections, the heat shrink is applied in the same manner we would use for welding. Uniform travel speed and standoff distance are important for getting consistent results. The inner pre-heat cones of the flame should be held off the surface to prevent scarring.

A full heat shrink on steel, as shown in Figure 10-16, will have the entire width of the shrink

heated to a low red or cherry red heat approx 1000–1300°F. In Figure 10-17, you can see I am straightening a stainless steel plate with a very light shrink.

Stainless steel is heated only until a deep brown is seen on the surface, less than 900°F (Figure 10-18). Because stainless steel is a poor heat conductor and can be damaged by overheating, we use a lower temperature when straightening stainless. Stainless steel expands about 1 1/2 times as much as steel, so less aggressive shrinking is necessary to get the same results. It's best to go easy and add more correction than to have to correct for overzealous application of shrinking. In Figure 10-18, you can see I am straightening stainless with a deep brown heat mark visible.

An important point is that once you have heated along a particular line with a full shrink, no useful

Figure 10-16: A full heat shrink on steel.

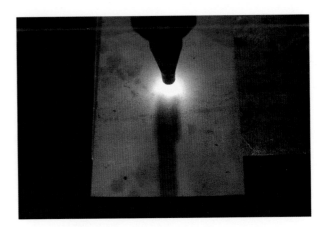

Figure 10-18: Barely a heat mark visible.

gain will happen by going over the exact same line, unless less than a full shrink was used. Normally if more correction is required, it is done along side of the original shrink line.

Here's a rough rule of thumb: If the plotted distortion 12 inches from the high spot is 1/16 or greater, the width of the heat shrink is equal to the material thickness. For stainless, reduce this by half for best results.

The heat shrink begins at an edge or in a single spot. For a full shrink on steel, the spot is heated until a dull red can be seen. The torch can now be moved along the length of the area to be straightened. It is important to maintain the temperature consistently as you move across the work. Remember, it's like reverse welding. Travel speed and standoff distance are as important as they are in welding. Weaving is recommended for a wider shrink area on thicker plates.

There are limits to how much correction you can get from heat shrinking. Typically you would see these limits near edges where plates are easily damaged in handling. It may not be possible to correct these types of damage without the use of aids like clamps or other mechanical aides. The effects of corrective heat shrinking will always be enhanced by the application of mechanical assistance. Weights and clamps will greatly magnify your heat shrinking results.

Cooling the Shrink

How we cool the heat shrink is almost as important as how we heat the work in the first place. Eventually the heat we input into the part will reach some equilibrium point if we leave it alone long enough. The effects of the heat shrink can be magnified by effective cooling of the heated metal. The cooling effectively freezes the upset from the heat shrink and gives us more correction per shrink than if no cooling is used.

For small jobs, a spray bottle of water or even a wet rag can be used (Figure 10-19). For larger jobs, a stream from a hose works the best (Figure 10-20).

Figure 10-19: Cooling the shrink.

Remember that an accurate assessment of the correction cannot be done until the part is cool enough to put your bare hand on (Figure 10-21). Additional shrinking should not be done in the same area until the part is cool enough to touch. Compressed air can be used to blow off excess water so you can see the gap between the straightedge and the part.

Cooling should start on the hottest part of the shrink first. This helps freeze the shrinks in their new contracted positions. Thin tubing and sheet should be cooled immediately after the heat is removed.

It is okay to apply several heat shrinks in one heating and cooling cycle, but they must be separated by cool metal to be effective. Heat shrinks

Figure 10-20: Using a hose.

Figure 10-21: When is the part cool?

should never be applied side by side without cooling unless two torches are used simultaneously for a wide heavy-duty shrink.

Figure 10-22 shows some of the other flame straightening plotting symbols I use. Most of these are pretty easy to understand. The main thing to remember is that these symbols represent what we see when the plate or part is examined. If you trust your eyes and measuring tools, the symbols are only a reminder of what needs to happen for a particular section.

A light dashed line represents a minimal application for a minor correction. A heavy-handed mark means a full power heat shrink to me. A stop shrink would mean the damage or distortion ended at that point. The same would apply to a gap shrink.

Straightening Shafts and Tubes

Shafts and tubes are straightened using the same techniques we use for plates. The easiest way to straighten a shaft is in V-blocks or, better yet, mounted between centers in a lathe with piped coolant available. How bent the part is determines what type of heat shrink we apply. With shafts or other cylindrical parts, we use either an axial or radial shrink. Excellent results can be achieved with this method. If you want to fuss around, you can easily get below .005 run out.

The axial shrink is the gentler of the two types of heat shrinks used on shafts and tubes (Figure 10-23). The radial shrink is for heavy duty distortion and should only be used if the axial type shrink does not produce the desired results. In the axial shrink, the heat is applied in the same direction as the length of the shaft. Normally these are very short shrinks because the apex or high points on shafts are very localized.

A dial indicator can be used more easily than straightedges to accurately locate the high points on a shaft (Figure 10-24). These points are plotted and marked in the same way we would mark a flat plate. Because we mark the high spots, all of your indicator readings will be plus readings. I use the center of the plus as my target when I apply the shrink.

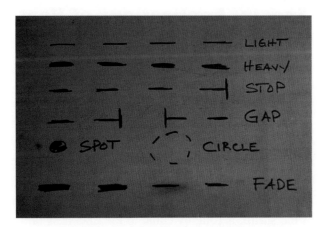

Figure 10-22: Flame straightening plotting symbols.

Figure 10-23: The axial shrink.

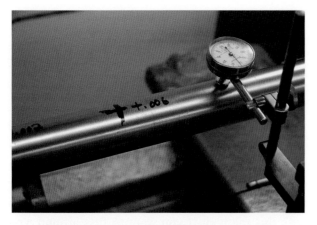

Figure 10-24: Locating high points on a shaft.

Figure 10-26: Controlling a radial shrink.

Figure 10-25: Finding the high spot.

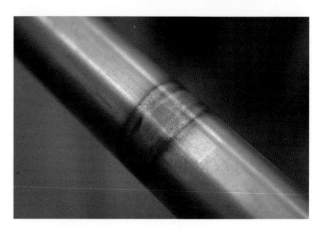

Figure 10-27: Working a stubborn bend.

The Aftermath of an Axial Shrink on a Solid Shaft.

The high spot in this example was .008 inches (Figure 10-25). When we apply the axial heat shrink to a shaft, we start a little before the high spot and continued the same amount beyond the apex. The maximum length for any axial heat shrink should be below two inches long. If more shrinking is needed, then a radial shrink should be used. The shaft in this example is 1 1/2 inch cold rolled steel.

For a radial shrink, the amount of shrinking is controlled by the radial distance that the shrink is carried around the circumference from the apex or high point of the distortion (Figure 10-26). For starting shrinks, about 20–30 degrees of radial arc would be shrunk. If the desired results are not achieved, more and more arc length is (Figure 10-27) added up to a maximum of about 120 degrees or arc. If this is still not enough, additional shrinks can be added along side, or mechanical aides can be used to work a stubborn bend.

Special Applications of Heat Shrinking

There are a few special cases worth mentioning here. The first is the high aspect ratio shrink. This is used for sections that have a greater depth than width, like rectangular bar or beam sections. Section depth limits how much correction we can get from normal shrinking of the apex of the distortion. For these deep sections, we need to involve more of the material. Remember the rule of thumb that the width of the shrink is roughly equal to the thickness or, in this case, the depth of the material. For these types of parts, we use a high aspect ratio or Vee shrink.

Figure 10-28: Connecting lines to form a Vee.

Figure 10-29: Heating and shrinking both sides of the Vee.

Figure 10-30: Starting at the widest part.

Figure 10-31: Using water and soluble oil to cool.

You can draw two lines separated by a distance approximately equal to the depth of the section, and then connect these lines together to form a Vee (Figure 10-28). This is the material we will shrink to correct this type of section. The Vee points away from the high spot of the correction. Typically both sides of the Vee are heated and shrunk (Figure 10-29).

When we heat this type of shrink, we start with the widest part and weave our way down to the point of the triangle (Figure 10-30). Mechanical aides such as clamps and weights can assist greatly in these cases. Cool the hottest part first. In Figure 10-31, I am using a spray bottle filled with water and a little soluble oil, which is the same coolant we use in the lathe.

Flame straightening in conjunction with mechanical aides allows straightening that would

otherwise be impossible with flame straightening alone. Typically the assists come from clamps, jacks weights, and even gravity.

Correcting Weldments

Weldments present some of the most demanding and complicated uses of flame straightening (Figure 10-32). Due to the interaction of the different parts and pieces, analyzing where and how to make the corrections can be tricky. Just remember the distortions are most likely caused from the welding that goes into an assembly. In Figure 10-33 you can see the bow in the tube caused from the welding on the opposite end. Target your corrections to counteract these effects. It is also helpful to make smaller corrections to sub-assemblies before they are incorporated into a larger structure

Figure 10-32: The challenge of weldments.

Figure 10-34: Producing a camber

Figure 10-33: The bow caused from welding.

Figure 10-35: Increasing the load carrying capacity.

where the distortions can become impossible to correct.

In this example, heat shrinks were applied to the underside of the bridge beams to produce a camber which increases the load carrying capacity (Figures 10-34 and 10-35). I would be willing to bet this was done with these beams sitting on sawhorses using gravity for assistance.

I have outlined the basic techniques of flame straightening in this chapter. The effective use of

this procedure can only be learned by trying it and carefully observing the results. The only real requirement is the willingness to try it. Flame straightening is not the answer to every distortion problem that faces a metal fabricator. It is only another tool in the tool bag of a competent metalworker. The importance of good design and the use of proper manufacturing techniques are at least as important as knowing how to correct for and repair the defects that are encountered in the welding shop.

Sheet Metal Shop

One of the old guys I learned a great deal from told me one time that a sheet metal worker's work was much more difficult than a machinist's work. I asked him to explain why he thought that was the case. Because I had worked on both sides of that argument, I figured I could offer an answer that would give him an appreciation of the difficulties faced in the machining trade. His reasoning was that a machinist started off with a block of material and slowly whittled it away until it was the desired shape and size. In contrast, the sheet metal worker has to first decide the starting blank size and predict the final size after working.

I gently pointed out that I thought this was a gross over-simplification by an old codger who had never turned the crank on any machine more accurate than a drill press. After the lights stopped spinning and the ringing stopped in my ear, I thought about his point and could see where he could get this idea.

Sheet metal work is a precision trade without a doubt. If you don't believe me, try welding a badly fit-up job in 24 gauge stainless steel and see how far you get without tight accuracy and precision fitup. Typically, sheet metal work is not as closely controlled as machine work, even though some designers and engineers might think it is. Production type work can be tuned and adjusted to very close limits, but only at the expense of many setup parts. One off

sheetmetal forming can be tricky to do accurately on the first part.

There are four basic types of sheet metal work: HVAC, Industrial, Electronic, and Aircraft and Automotive. They all overlap somewhat and the techniques learned in one class of work are easily applied to any of the others.

The first type, HVAC, involves heating, ventilation, and air conditioning, as the name suggests. HVAC work is characterized by lots of special seams and lock joints with shape transitions, very little welding, and lots of duct tape and self-tapping screws.

Industrial sheet metal generally covers chemical and food processing equipment, conveyors, tanks,

Figure 11-1: Sign in front of the sheet metal shop at Hunters Point Naval Shipyard.

and vessels made from sheet metal. This class of sheet metal work is in many cases liquid or pressure tight, sometimes in special corrosion-resistant materials with lots of welding, plumbing, and polishing.

Electronic sheet metal includes the sheet metal formed into the chassis and brackets used in computers and other enclosed electronic equipment. Electronic sheet metal has lots of punched or laser cut holes and cutouts, with inserted hardware, high accuracy, and minimal welding. Most forming is done in press brakes, sometimes using special tooling

Aircraft and automotive work covers sheet metal formed into compound shapes or warped three–dimensional surfaces. Much of this compound work involves more art than science at the individual worker level. It is easily the most complex work, where a solid understanding of material behavior and welding skill is required. It is also characterized by large amounts of hand work.

Obviously this is somewhat of a generalization on my part with many subsets of each basic type in actual industry, but it can be said that each of the basic types has its own particular set of problems and challenges. Keeping to the trend I think I have started in this book, I will focus on tips and tricks that cross over between the different types and have use to anyone involved in sheet metal work in the jobbing shop setting.

My definition of sheet metal is sheet material that is named by a gauge number or less than .188" thick measured decimally. For most sheet metal applications, it is less confusing for the designer and the shop personnel if decimal material thickness with tolerances are called out on engineering drawings. I would classify material heavier than .188 as plate work, even though the same calculations and techniques are used with sheet metal.

All sheet metal work starts with layout and planning, more than any other trade, because we must anticipate and predict how the material behaves before we touch tool to metal. Patterns and templates are commonplace. With the increasing use of computer pattern development and computer-controlled, high speed laser and waterjet cutting,

the need for expert pattern development is changing for some shops. During my early years, before computers, I found pattern development to be challenging and fun.

Classical pattern drafting is outside the scope of this book and is covered in excellent detail in many readily available books. I have listed several in the recommended reading section in the back of this book. So, what kind layout are we talking about? To me layout is the ability to plan and predict how the material will behave before you do it for accurate results.

Layout Work

Skip the bluing unless you really enjoy the smell. Dykem is for the days gone by before good carbide scribers. Press a little harder with your scribe. Save the drying time and the inevitable cleanup. In a pinch, you can use a little Sharpie marker to provide a better contrast on a light-colored surface.

A Sharpie marker is used in Figure 11-2 to bring out detail on a difficult surface. Instead of coating the entire surface with bluing, just cover the necessary areas of detail (Figure 11-3).

There are several tools that can be used to mark a line parallel to another edge (Figure 11-4). For small dimensions less than an inch, a modified pair of dividers with a guiding edge and a scribing edge coupled with a screw adjustment is fast and accurate (Figure 11-5). The friction pivot type is quite lame and always seems to move just when you don't want it to move.

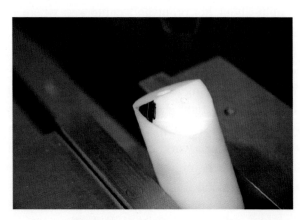

Figure 11-2. Bringing out detail.

Figure 11-3: Covering just the areas of detail.

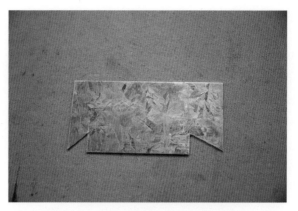

Figure 11-6: A quick scratch template.

Make a quick scratch template from sheet metal for repetitive layouts (Figures 11-6 and 11-7). These two are 24 ga galvanized sheet metal. You can mark bend lines and hole centers for repetitive layouts.

I worked in one shop where the whole place ran on sheet metal templates (Figure 11-8). They were all numbered and stored in a special rack.

Everything you needed to do the job could be found directly on the template. Hole locations are punched through the small holes and bend locations are marked in the notches. One of the best features of a template is that it is a physical representation and gage of the part, not a scaled picture or facsimile that requires mental interpretation to determine if it's correct.

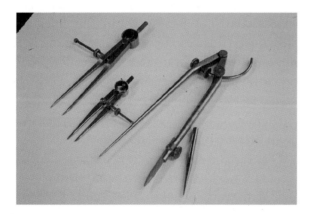

Figure 11-4: Handy tools for marking parallel lines.

Figure 11-7: Using sheet metal for a template.

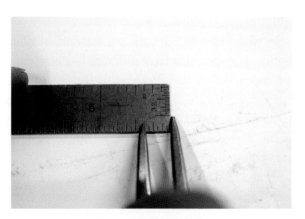

Figure 11-5: Measuring less than an inch.

Figure 11-8: Identifying sheet metal templates.

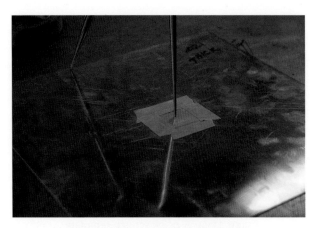

Figure 11-9: Using a sacrificial plate.

Tape the center point for dividers on work that will need special finishing, or extreme shaping or forming. An alternate method is to use a sacrificial plate to put your layout and center punch marks on (Figure 11-9). The center divot from your dividers or scriber can be a point of failure and is a pain to polish out on high finishes.

When doing your bend layout, it is customary to indicate the direction of the bend so mistakes aren't made later at the brake. Only scribe or punch mark the inside or compression side of the bend. Bends indicated as "down 90" should be marked with felt pen only on the outside or tension side as a reminder which direction to bend. Sometimes the sheets are too large to easily flip over to properly mark the inside. Marking on the inside prevents tearing or cracking out of the scribe or punch marks (Fig 11-10).

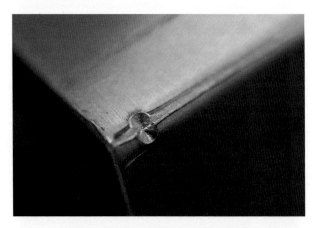

Figure 11-10: Marking on the inside.

Blank Length Calculations

Blank length calculations are one of the deep mysteries surrounding sheet metal work. The ability to mathematically predict in every situation the amount of material used during the forming of the material is one the cornerstones of sheet metal work. This little bit of mystery material is called the "Bend Allowance." It is also the cause of much frustration.

I would like to set the record straight if possible. The bend allowance is not something you can predict with absolute accuracy—period. There are so many factors involved in calculating bend allowances that affect the final outcome of the bend that it should always be though of as an approximation as opposed to an absolute. If the sheet metal work is toleranced properly, the difference between what can be predicted and what can be achieved is the tolerance.

Admittedly you can get pretty close with calculations and bend allowance formulas. Predictable results come from predictable methods and tools. Suppose your shop bends a lot of .048 thick cold rolled steel. Chances are all the bend allowances, compensation factors, and tooling has been worked out over enough work so that very predictable results can be achieved. Now in that same shop, toss in a piece of .125 thick soft copper sheet and all bets are out the window.

What I am heading for here is the statement that if accurate work or extremely tight tolerances are required, then test bends will be required. There is no getting around this. Material variation alone will account for a large part of the uncertainly in predicting bend allowances.

So what we are left with are systems of blank length calculation that have different levels of accuracy, repeatability, and complexity. Repeatability is what we really care about in bend allowance calculations. The ability to reproduce your forming results, day after day and material lot to material lot, is where the rubber meets the road.

Figure 11-11a shows the basic bend configurations that come up in the sheet metal shop. There

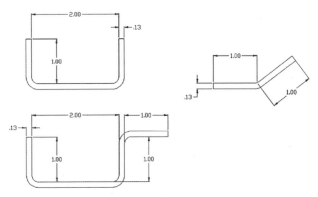

Figure 11-11a: Basic bend configurations.

are two things we need to produce a bend in sheet metal. 1) What length of sheet metal is needed to produce the desired bend configuration? 2) Where do I place the bends on that blank? For this explanation, I am assuming standard bends produced from standard tooling for this material thickness.

The first blank length calculation method is the easiest. The bends don't take much material; therefore, the assumption in the calculation is they are sharp. What sharp means is no radius is accounted for in the math. The calculation is quite simple and for a large majority of the work has more than enough accuracy. To calculate the blank length, we simply add the bend to bend lengths. In other words, we add the lengths on the inside of the shape. For the upper left example in Figure 11-11, the blank length is 1.00 + 2.00 + 1.00 = 4.00. This is the length of sheet metal we cut to produce this channel. Accuracy in this example is easily +/− .06.

For the lower left shape in Figure 11-11, we still add the lengths between the bends. One of the bends is on the opposite side from the two others, but as long as we add the bend to bend lengths, we will get the proper length. Thus, 1.00 + 2.00 + 1.00 + 1.00 = 5.00.

For the upper right shape in Figure 11-11, it's dog simple: 1.00 + 1.00 = 2.0.

Now we will have to locate the bends on the blank lengths we just calculated. With the simple

method true to its name, we just locate the bend lines using the dimensions off the drawing. For our first example, the channel in Figure 11-11, we simply take our 4.00 long blank and mark a line in from the end 1.00 on each end, then bend. Done deal. For the remaining examples, they would have their bend lines marked in the same way.

That was one set of examples for several bends on non-specified material other than the thickness. For more predictable results, and I mean better than +/− .06, there are more considerations put into the calculations. Here are a few things that affect the results of bent sheet metal.

Material type and temper. Harder materials have greater bend allowances. Or another way of putting it, more material is used per bend in hard materials.

Angle of bends. Bends of just a few degrees use less material than acute bends or bends past ninety degrees.

Thickness of material to be bent. The blank length calculation for accurate bends takes into consideration the neutral axis of the material. The neutral axis is the theoretical line somewhere near the center of the material thickness that is neither stretched nor compressed. The neutral axis does not change length during forming, unlike the inside and outside of each bend. On thicker material, this becomes a significant addition to the math.

Type of forming tooling. The tooling used to make the bends has an effect on bend allowances. What is calculated for bend allowances with one set of tooling does not apply to another.

You can see now why there is so much confusion around this subject of bend allowance. Everybody has their own way of doing it based on personal experience and how they were taught. Computer-controlled press brake manufacturers have incorporated their own methods for determining bend allowances and built them into the controls of the machine. As I said earlier, these calculations are not an exact science so they build some adjustability into the machine controls that allows the operators to adjust for the exact conditions they are experiencing. So the next time someone tells you bend allowance calculations are an exact science, ask them why they have this adjustability in the machines. Milling machines are not adjustable like this. They cannot be operator adjusted to cut .001 one day and .010 the next.

The next method for figuring bend allowances and blank lengths is considerably more accurate and reproducible. The first step is to take the material hardness into consideration. This is the first step because the hardness of the material determines approximately where the neutral axis lies, which affects its overall length. All the neutral axis numbers are located from the inside surface of the material or the inside (Figure 11-11b). Here are three numbers to begin.

Soft materials. For copper, soft aluminum, lead, and gold, the neutral axis is approximately .55 of the material thickness from the inside surface—in this case, a tiny bit past halfway. Another way to write this is .55T

Medium half hard materials. Neutral Axis = .64T. These include materials like soft annealed or killed steel, non heat treated aluminum, half hard copper, and brass.

Hard materials. Neutral axis = .71T, and includes materials like 304/316 Stainless, cold rolled steel, phosphor bronze, and spring temper steel and stainless.

Now for the next part of the problem: the inside radius. Inside radius is calculated either from the bend produced by standard tooling or from the part drawing. Here's the rub. Now that the designer has gone and specified a radius, you are obligated at least to try to produce it to specification. Hopefully the designer has specified the radius produced from normal tooling and not something too much larger or smaller. We're talking vanilla-flavored, 90-degree air bends here produced in a press brakes. Any oddball stuff has to be worked out anyway.

The issue with the inside radius is that it may require special tools to produce. The sheet metal operator has a limited range of radii that can be produced from the normal tools for the material thickness and angle of bend.

For normal standard bends in Vee-type dies, you calculate the inside radius by starting with the material thickness and finding the correct tooling. Once the tooling has been selected, the inside radius can be approximated, which will eventually lead us into the blank length calculation. This all sounds much harder than it really is, so don't worry. That simple method is starting to look pretty good right about now.

So we have three possible conditions for the inside radius and their relative ease of producing.

1. Sharp. No calculation required for bend allowance. Easy.
2. From standard tooling. Pretty easy.
3. Non-standard specified radii. Easy-to-difficult.

In a nutshell, the inside radius from standard tooling is a percentage of the width of a Vee die 8 times the material thickness for material up to .188 thick. Putting that sentence into equation form we get,

$$R = (8T).156$$

For example, if the material is .125 thick, then

$$R = 8 \times .125 \times .156 = .156 \text{ inside radius}$$

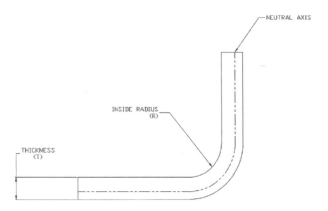

Figure 11-11b: Considering material hardness.

Following through the formula for bend allowance for a 90-degree bend is:

$$\text{Bend allowance (BA)} = \text{(Neutral axis)} \times \text{(Thickness)} + (1.57 * \text{Radius})$$

For example, for .125 cold rolled steel:

$$(.71) \times .125 + (1.57 \times .156)$$
$$= .3337 \text{ Bend allowance}$$

This means each 90-degree bend uses .3337 inches of material per bend. The next step is to add the bend allowances to the straight sections of material to arrive finally at our official blank length.

Because we know the inside radius of the example is .156, we subtract that from the inside lengths from Figure 11-11. Adding it all together,

$$(1.00 - .156) + (2.00 - .313)$$
$$+ (1.00 + .156) + (.3337 \times 2)$$
$$= 4.043 \text{ (Figure 11-11c)}$$

Applying this formula to our original example, we get a blank length of 4.0434 (Figure 11-11c).

All of these calculations, just to figure the blank length, seems like a lot of work. It is, but if you want to do accurate work and have reproducible results, you at least have to learn how to do the math. All of this illustrates the theoretical method for blank length math.

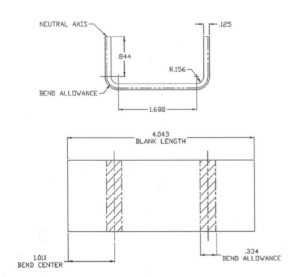

Figure 11-11c: Applying the formula.

I'm sure you noticed, but the blank length from the "accurate" method is only .043 longer than the length we figured using the fast method. The question is: how close do you need your bends to be? With all that calculating, these are still just numbers. Until you make a part, you are only living in the theoretical world. You will be happy to hear that there is an easer way to calculate blank lengths.

As I pointed out, the accurate blank is only .043 longer than the fast blank. Why not somehow combine these two and have an easier way of getting accurate results. You can combine them and not be stuck in the theoretical world. The way we handle it on the shop floor is to make test bends.

Test bends are the only real way of obtaining super accurate results anyway, so why not start there. If you just want the bend allowance, cut a blank (of arbitrary length) of the exact material you intend to use for your part and make a bend. Be sure to measure this test piece before you make your bend. After forming, measure it again and determine the actual loss for one bend. If you measure accurately you now have the best real-world non-theoretical example possible. The key things to remember using the test bend method are the material must come from the same lot and sheet as your part. The only other factor of significance with the test strip method is the test strip should have the same length of bend as the real part will. If the part has three inches of bend, then the test strip should have three inches also.

Charts of bend allowances for different materials are readily available to sheet metal workers. *Machinery's Handbook* has an excellent section on this very subject with look-up tables for allowances. The bottom line is that most of the time you can skip the heavy math and either make test bends (recommended) or use look-up tables and still have to make test bends. The choice seems simple enough to me.

Notching patterns and angle guides. We use a simple template to set up the power notcher for repeat cuts (Figure 11-12). Figure 11-13 shows corner notches for cabinets with recessed doors.

Figure 11-12: Setting up the power notcher.

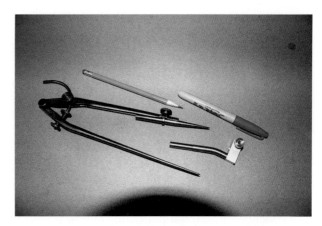

Figure 11-15: Large dividers and trammels.

This special notch allows the forming of a formed mitered corner (Figure 11-14).

For sheet metal layout of average size, you will need a set of large dividers and trammels (Figure 11-15). Be sure to buy some that can hold a regular pencil. These great dividers (Starrett®) are expensive, but will sweep an 18-inch circle with a scribe point or a pencil. The bar type is superior to the dividers made from flat stock. The flat material twists when you lay into it trying to get a decent scribe going.

The little home-made device seen in Figure 11-16 clamps a faithful Sharpie marker in one leg of the dividers or trammel. Figure 11-17 shows another

Figure 11-13: Corner notches.

Figure 11-16: Clamping a Sharpie marker.

Figure 11-14: A mitered corner.

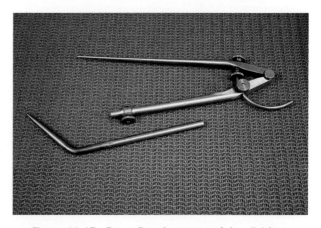

Figure 11-17: Extending the range of the dividers.

home-made tool that extends the range of these dividers even farther.

Patterns

Pattern paper is typically used for initially developing shapes other than very simple patterns. Using paper has several advantages, particularly when the pattern will be used several times or the metal that the pattern will be transferred to is expensive, such as copper, silver, or stainless steel. For sheet metal pattern drafting, a medium-to-heavyweight paper is used (Figure 11-18), if the pattern will be used in the shop or the paper pattern will actually be formed to check forming, tooling, and finished sizes. It is acceptable to lay out the pattern directly on the material, but a paper pattern eliminates all the construction scribe lines and arc centers that are normally required to develop a sheet metal pattern.

If a more durable pattern is required for marking off many transitions, light gauge sheet metal can be used. We use 26ga galvanized sheet for durable patterns. It is soft and easy to cut and the plating keeps the patterns from rusting. The original pattern is typically still done on paper. If graphic methods are being used, a sheet larger than the actual metal blank are required.

Pattern paper should be tough and fairly thick, with enough width to make an entire pattern without tacking together smaller sheets. At some of the home supply centers, red paper rolls with decent weight can be purchased. They are used to protect floors during construction. This paper is suitable for sheet metal patterns. It has enough thickness that scribe lines can be easily transferred to the metal blanks without the paper moving or the scriber slipping off the edge. Our rolls are 48-inches wide. This paper has quite a few other uses in the shop for protecting surfaces and parts.

I like to secure the paper directly to my workbench with spring hand clamps (Figure 11-19). Duct tape works also but doesn't allow you to re-position the paper as easily.

Your trammels should have a wide adjustment range for the size bar they fit. They should also easily hold a pencil. Fine adjustments are made by turning the eccentrically ground points. The type in Figure 11-20 fits a wide range of bar sizes, including rounds. For very long radii and arcs layouts, spin off some MIG wire and use that for your bar.

Figure 11-19: Securing paper with spring hand clamps.

Figure 11-18: Sheet metal pattern drafting.

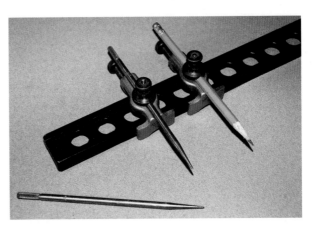

Figure 11-20: A trammel with a wide adjustment range.

The "Yank Method"

The old sheet metal crustacean I worked with taught me a neat method for transition development. He was born and raised in New Zealand and to him all Americans were "Yanks." He called this particular method of transition development the "Yank method." It was meant as derogatory because it is not considered a "pure" method of pattern development when compared to the more traditional graphical methods we have all seen. I find it funny that a Kiwi hiding out in Yank-land had to teach the so-called Yank method to an American. However unpure the method was, it didn't stop him from using it.

Also called the short or roll method, it uses a three-dimensional physical model of the transition to develop the flat pattern for the sheet metal. The method involves connecting two end shapes of almost any transition together in their desired relationships, angle, distance, etc., and combining them on a common axis. This model is either rolled directly on the sheet metal or rolled on pattern paper and re-used as many times as necessary. The path of the end shapes are traced as it rolls, creating the proper stretch-out and angular relationships in the flat pattern. The beauty and advantage of this method is easy to see it you have ever had to measure an odd transition in the field and get accurate results.

Figure 11-21 shows a table mockup of a typical transition you might encounter in the field. (The end shapes are irrelevant for the demonstration.) In this case, we will do a simple rectangle-to-round (Fig 11-22). This set of conditions would be a real pain in the neck to measure and do a decent job of it. I was lucky enough to set this up on my welding table, which happens to have a rectangular hole in the center. Try doing that under a machine that happens to be running while you're trying to take decent measurements. This method works with almost any shape of end terminations. Transitions from diamonds to ellipses are possible and just about anything you can think of in between. The beauty of this trick is that no measuring is necessary to develop the pattern. The first step is to cut

Figure 11-21: Table mockup for a typical transition.

accurate filler plates for each of the ends. They should fit the openings in the same way you want the transition to fit. If the transition will fit the outside, then the plate should be that exact diameter.

Figure 11-23 shows the physical "Yank" mockup of the transition I am demonstrating. The two ends are locked in their exact proper relationship

Figure 11-22: Preparing for a rectangle to round.

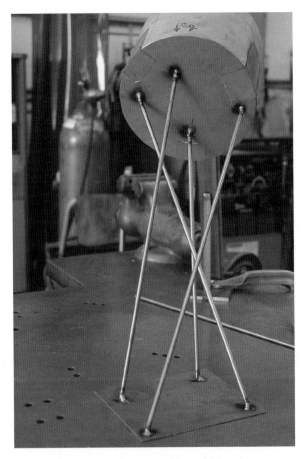

Figure 11-23: The physical "Yank" mock-up.

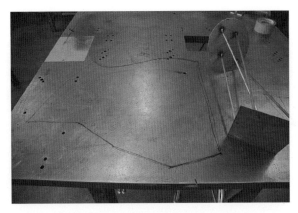

Figure 11-24: An asymmetric example.

Figure 11-25: Tracing along the edges.

by the welded rods and not subject to eyeballing or sloppy measuring errors. The mockup should have enough strength to be handled without the ends moving or breaking off. I like to use 1/4-diameter rod because it's cheap and easy to cut with a small pair of bolt cutters in the field.

I always flop the mockup around through a complete revolution just to get an idea how big of a piece of paper I will need to trace the entire pattern. The example in Figure 11-24 is asymmetric so we will need the full revolution of the pattern. For symmetrical transitions, you can just develop half if you prefer.

Trace along the edges of the mockup as it is slowly rolled through a complete 360-degree rotation (Figure 11-25). Mark the end points of the rectangle when the edge sits flat on the paper. I also like to mark the circle at the exact spot it is touching the paper when the straight edges of the rectangle are resting on the paper. This gives me an idea

where the rounding bends will be placed. It's sometimes tricky to determine where the bends will be placed; therefore, it's a good idea to work it out on pattern paper before you switch to sheet metal. If the mockup slips or is bumped while you are rolling it, you should start over.

In Figure 11-26, the mockup has been rolled through a complete revolution and the paper pattern

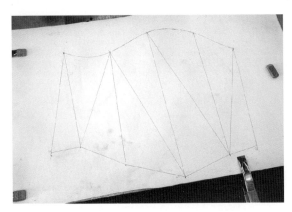

Figure 11-26: A completed revolution.

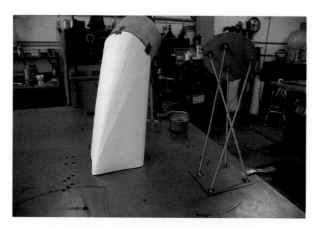

Figure 11-27: Fitting the pattern to the setup.

developed. I have marked where the rounding bends will start and end. Remember, this is the inside of the transition. I have accidentally bent a couple inside out because I was so anxious to see the transition formed up.

In Figure 11-27, you see the paper pattern fits the test setup pretty well. Paper doesn't hold its shape that well, but it will give you an excellent reading on the success of your pattern development.

For these kinds of transition forming jobs, the box and pan brake is a wonderful tool (Figure 11-28). It is much quicker to set up than the press brake and makes these types of multiple rounding bends with a single intersection with ease. The top clamping beam is set back further than the normal material thickness to more like 1 1/2 − 2 material thicknesses.

Box and Pan Brake

Box and pan brakes are extremely versatile machines if you know a few tricks to make them talk. These machines lack dedicated back gages to do repeat forming work like press brakes. A simple back gage can be bent up from a piece of sheet metal thinner than the part you intend to form (Figure 11-29). You can just see it in this picture under the finger. These back gages are clamped to the machine frame from the rear. Box and pan brakes are faster to set up than press brakes for a few odd brackets or multiple angle bends. Box and pan brakes are my favorite machine for bending springs clips and other small multi-angle fabrications because of the ease of setup.

For multiple bends, you may find that you need to gage off the front apron. This can be done several ways. One way that saves on layout for multiple parts is to index off a reference bar or stop on the front of the machine (Figure 11-30). The box and pan brake has the ability to make opposite side bends quite close together without the special tooling that would be required in a press brake.

For those painful snipping jobs, you can get a better purchase on your snips if you clamp them in the vise. A short length of tubing completes the setup for making short work of an otherwise painful operation. The snips shown in Figure 11-31 are 25 years old favorites made by Bahco tool. These are some of the toughest snips I have ever used.

Figure 11-28: A box and pan brake.

Figure 11-29: A simple back gage.

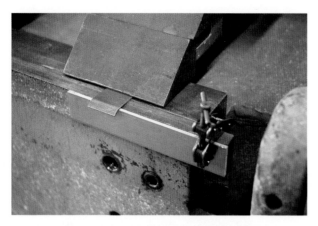

Figure 11-30: Indexing a reference bar.

When forming pans and trays in the sheet metal shop, provide a corner relief hole (Figure 11-32). The thicker the material, the more important this becomes. It's easy to remember the size of the relief hole because its diameter is twice the material thickness. Use this corner detail if the formed corner will be welded. The corner will come together snugly, allowing for a nice clean seal weld.

A typical problem with common trays and pans is oil canning across the large flat surface (Figure 11-33). There are a few causes of this annoying easy-to-cure problem. The most common cause of oil canning is the corner welding. If the corners are to be welded, be sure the flanges fit closely so minimum heat input is required. Back up the corner weld with a heat sink block of copper or aluminum (Figure 11-34). This will draw off some of the

distorting heat from the joint. If you still have problems, you can use a hammer and dolly in the corner to peen the weld to stretch it a little (Figure 11-35). If your tray or pan was flat before you welded the corners, chances are the welding caused the problem.

Use a corner notch detail like the one in Figure 11-36 for formed trays and pans when the corner will not be welded. Using this relief pattern, the corners will not interfere with each other during forming. This makes for a nice straight flange without bulging corners.

For non-welded production trays or covers, the entire corner can be lopped off at a 45-degree angle for super simple corner preparation (Figure 11-37). Be sure to clip the corner back far enough to clear the press brake tooling (Figure 11-38). Use this corner detail so you don't have to make up a special length top punch, thereby saving setup time.

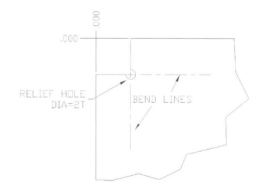

Figure 11-32: Providing a corner relief hole.

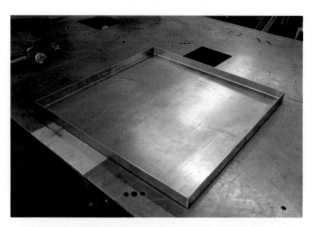

Figure 11-31: Some tough snips!

Figure 11-33: Oil canning is a typical problem.

Figure 11-34: Backing up the corner weld.

Figure 11-36: A corner notch detail.

The layout for the corner notch in Figure 11-39 is the flange height, plus one half the width of the top punch plus 1/16. This is the distance from the corner of the sheet for the 45-degree clip.

The inside radius of a bend formed in the press brake is dictated by the width of the lower die opening (Figure 11-40). Many people mistakenly believe that the upper die or punch nose radius controls the formed radius. This is not the case as you can see in Figure 11-41. The only difference in the two bends was the lower die Vee width If you want to change the inside radius, change the lower Vee die width.

The upper punch nose radius should be 20% less than the calculated theoretical inside radius for the desired bend. Truth be told, the inside curvature of a formed bend is not even round. It's actually elliptical in form. This is just one reason the blank length calculation formulas are really only theoretical approximations and do not always produce accurate results.

The only difference in these two air bends was the width of the lower die opening (Figure 11-42). The smaller radius was formed in a Vee die that was 1.125 and the larger radius was a 2.00 Vee opening. Upper punch was the same for both bends.

If your bends must be super accurate, you will have to make test bends. There is no getting around this except pure luck. With practice and consistent materials, you can get pretty close, but there is no

Figure 11-37: Lopping off the corner.

Figure 11-35: Peening the weld to stretch it.

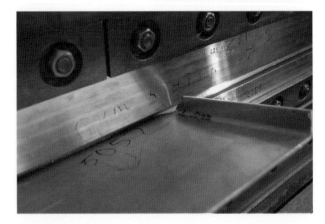

Figure 11-38: Clearing the press brake tooling.

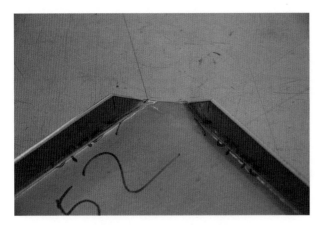

Figure 11-39: A corner notch.

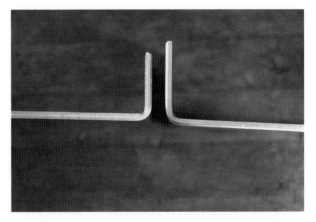

Figure 11-42: Comparing the width of the lower die opening.

substitute for test bends in the actual production material.

Standard air bending Vee dies want to have an opening eight times the material thickness. This ratio increases for materials over 3/8 thick where a die opening of ten times the material thickness is generally used. For instance, 1/8 thick material

Figure 11-40: The inside radius of a bend.

Figure 11-41: Comparing the lower Vee die width.

would use a 1-inch Vee opening lower die. The basic formula for calculating the approximate inside bend radius in steel is,

$$(MT8) \times .156$$

that is, material thickness multiplied by 8 multiplied by 5/32. This calculation gives us an inside radius of 5/32 for 1/8-thick material formed in standard air bending dies. This formula is an approximation for medium hardness materials like cold or hot rolled steel. Materials that are harder or softer than steel will not conform to this rule of thumb. Harder, stiffer materials will have larger radii and softer materials will have much smaller radii.

For some reason, many designers believe it's important to specify the bend radius for a sheet metal or plate forming operation even when it has no impact on the design (Fig 11-43). The only constructive

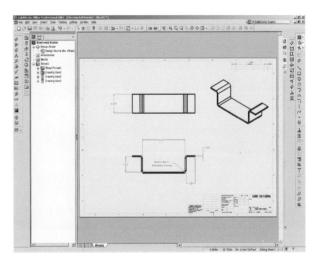

Figure 11-43: Specifying the bend radius.

Figure 11-44: Changing the internal radius.

Figure 11-46: Changing the overall dimension.

advice I would offer is if the bend radius is not important, leave it off the drawing. An alternative would be to apply a liberal tolerance to any specified bend radius, or note the radius as, "From standard tooling." Many hours of frustration in the shop have resulted from a bend radius callout that was difficult or impossible to produce because it fell under the title block tolerances. The worst is when quizzed, the draftsman or designer often says, "Oh that, it's not really important—just any bend." Arrrgggg!

You can change the internal radius after forming by over-bending the angle and then back bending back to 90 degrees (Fig 11-44). The material is pulled a little further and forms a slightly smaller radius. Because of this little extra forming, the material is a little stiffer. Therefore, when the back bending operation takes place, the net gain is a slightly smaller bend radius—a handy trick when confronted

by a bend radius with a tolerance on a drawing. It's a small radius change, but sometimes necessary.

This same trick can be applied to adjusting the distance between two bends. It is seen in bracket bending involving a closely-controlled distance between legs. This distance can be adjusted slightly by over-bending a few degrees and then back bending to vary the overall width by small amounts. Figure 11-45 shows the channel dimension after the first two bends.

In Figure 11-46, the overall dimension has changed after re-striking the flanges and slightly over-bending the angle beyond 90 degrees.

Figure 11-47 shows the channel after back bending the flanges and returning them to 90 degrees. The net change in the overall dimension was .010 inches. More change can be gained in heavier materials and more severe over-bending. Experiment

Figure 11-45: The channel dimension after two bends.

Figure 11-47: Further changes to the overall dimension.

Figure 11-48: Altering bend positions by pushing.

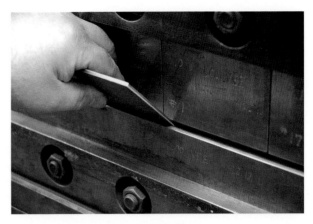

Figure 11-50: Rough setting the depth.

with this trick the next time you have one piece of material and you just formed the legs a little bit out of tolerance.

Bend positions can be altered slightly by pushing (Fig 11-48) or pulling (Fig 11-49) on the part as the dies close on a re-striking. Offsetting the lower die has a similar effect but is not always possible or practical on every machine. You may have to correct for slight over-bending. But if you only have one shot at your forming, this can be a useful trick to adjust a very small amount. Beware: this can also change your flange height at the same time.

Much of the forming work I learned on was using expensive, thick materials. Cutting another piece of $10-per-inch flat bar or using up miles for extra test bends was not acceptable. You end up finding a way to make small corrections for the

inevitable errors in blank length and bend allowance figuring.

You can quickly rough set the depth of a set of air bending dies for a 90-degree bend by setting the bottom position with a piece of the same material to be formed (Fig 11-50). Jog the ram down until the scrap part is almost clamped. This will be pretty close to a 90-degree bend.

When forming tapered ends, keep the end of the taper away from the intersection of the bend radius. You can see in Figure 11-51 that when the taper intersects the bend line, there is a distortion which makes for sloppy fitup later on. If you increase or decrease the taper amount to move it away from the bend radius, you will get better results (Figure 11-52).

An easy way to add strength and stiffness to a bracket is to form gussets directly into the brackets

Figure 11-49: Altering bend positions by pulling.

Figure 11-51: Distortion that makes for sloppy fitup.

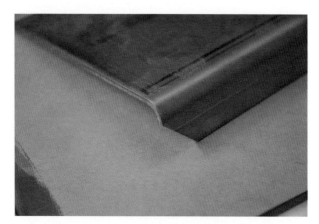

Figure 11-52: Changing the taper amount.

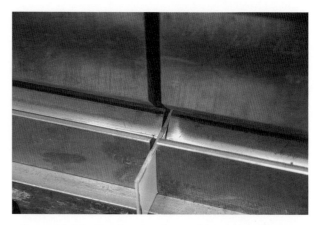

Figure 11-54: Spacing the dies and inserting a flat bar.

(Figure 11-53). These are created by spacing the bottom and top dies, then inserting a flat bar between the lower die (Figure 11-54). The depth of the gusset is controlled by how much it protrudes into the forming area. A word of caution: go easy. A material thickness into the forming area is all that's required. The forming bar should be rounded and smooth to prevent shearing the part and the upper die spacing needs a little extra room for the metal to expand outward along the bend. If you're forming a bunch of brackets that need a little extra stiffness, this is a great trick.

Here is a shot of some shop-made gooseneck dies (Figure 11-55). You just never seem to have just the correct die for some of those tightly-formed channels (Figure 11-56). These were made from a couple of heavy angles in the fabrication shop in about an hour.

One trick that's handy for forming deep channels and other forms is to use a back bending technique

(Figure 11-57). This is not the first choice on the list for production bending, but it can get you out of a jam for a couple of pieces when you don't have the specialized tooling to produce them.

The first step is to put a mild bend exactly in the center between the two legs (Figure 11-58). This bend should be something like 15–20 degrees,

Figure 11-55: Shop-made gooseneck dies.

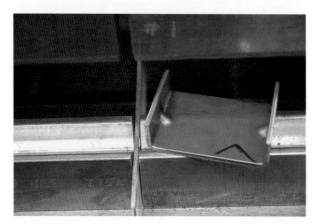

Figure 11-53: Forming gussets directly into the brackets.

Figure 11-56: Finding the correct die.

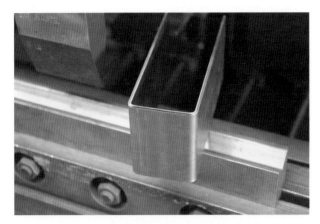

Figure 11-57: A back bending technique.

Figure 11-60: Aligning and bending back.

depending on the length of the legs. The 90-degree legs are then formed, making a W shape (Figure 11-59). The first bend allows the 90-degree bends to clear the machine.

The last step is to carefully align on the exact center between the two legs again, then back bend the W flat again (Figure 11-60). I find it helpful to

mark the centerline on both sides of the blank before I start.

Sandwich perforated sheets that have a linear pattern or holes that will distort when rolling (Figure 11-61). This method also works well for sheets that already have holes or cutouts in them (Figure 11-62). Sandwich the work piece between

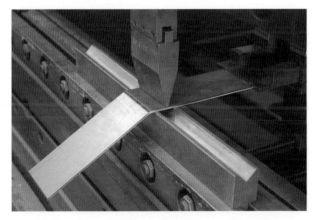

Figure 11-58: Putting a mild bend in the center.

Figure 11-61: Sandwiching perforated sheets.

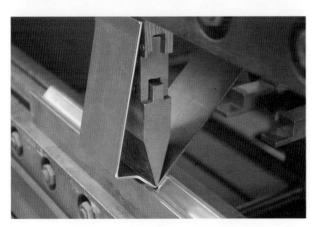

Figure 11-59: Making a W shape.

Figure 11-62: Sheets that already have holes and cutouts.

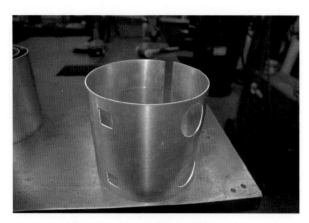

Figure 11-63: A facet-free cylinder.

Figure 11-65: Forming the legs closely together.

two thinner sheets, then roll normally. The results will be a facet-free cylinder (Figure 11-63). The inner sandwich sheet needs to be a little shorter than the part sheet; otherwise, there will be an overlap problem as the cylinder closes.

Here is a special punch for the press brake, sometimes called a widow punch (Figure 11-64). This cutout allows the legs of a channel to be formed closely together (Figure 11-65). It is very limited on applied tonnage, but handy for those impossible bends that seem to crop up. This punch was shop-made for a special shackle-forming job.

When making large cuts on the shear, a simple trick to save your back is to slip the width of the sheet on top of the back gage to make your roughing cuts (Figure 11-66). This keeps the heavy drop from falling down into the bottom of the drop chute behind the shear. You can now go around the back of the shear and slip the large drop back up

onto the deck of the shear for accurate squaring. Believe me, if you have to shear stacks of 1/4-material all day long, this will save your spine some serious abuse and probably several trips to the back cracker's couch. This trick may not work on all shears because of the height of the back gage and the width between the housings. Set the gage at approximately half of the length of the drop. Doing so allows it to balance on the back gage and let you slip it back onto the shear deck.

Use front stops for repetitive cuts (Figure 11-67). This keeps the sheet on the operator side of the machine for faster loading and unloading for large pieces. Once set, these are very accurate; they're not affected by sheet droop like the back gage. A quick way to set the front stops accurately is to shear a strip the exact length you need. Then jog the shear so the top blade is just crossing the bottom right in line with the front stop. You

Figure 11-64: A widow punch.

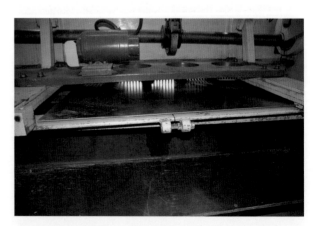

Figure 11-66: Making large cuts on the shear.

Figure 11-67: Using front stops for repetitive cuts.

Figure 11-69: Blanking when shearing small pieces.

can now butt the strip against the blade and run the front stop into position. If you have a mechanical shear, you can cut individual strips of scrap material and measure the length to adjust the stops accurately. This trick, combined with the other trick shown in (Figure 11-66) using the back gage to support large drops, can save you a lot of hard work.

Always strip (Figure 11-68), then blank (Figure 11-69) when shearing small pieces. The order of shearing has a great affect on accuracy and speed. Long narrow strips are hard to handle and cut accurately. The narrower and longer the strip, typically the more twist it has.

Here is a simple little trick to get your parts tacked together. Sometimes it's a chore to hold both the part and the TIG torch, and then on top of that step on the pedal to get a part tacked. If you put one part out of alignment in one axis (Figure 11-70),

you just align the corner of the joint for tacking; it makes this juggling act a little simpler. In this case, the fitup is inside corner to inside corner. All we need to think about is keeping a little downward pressure and the parts will stay in alignment.

After you get a small fusion tack (Figure 11-71), the top sheet can be rotated carefully into correct

Figure 11-70: Putting a part out of alignment.

Figure 11-68: Stripping when shearing small pieces.

Figure 11-71: A small fusion tack.

alignment for the next tack. All this is done one handed.

Forming and Layout of Cones

The cone is one of the basic shapes encountered in sheet metal. Years ago, somebody gave me a little sheet metal handbook that had a great little method for laying out truncated cones, which is how all fabricated cones are made. This is the method I use exclusively now.

The diagram in Figure 11-72 shows cone dimensions and naming conventions. Eventually you won't need the diagram. You may even make a cool little spreadsheet that you plug your numbers into to get the output.

Figure 11-72 shows the actual flat pattern. The secret to the method's accuracy is the chord dimension (Figure 11-73). Because we are using the chord of the arc segment, we bypass the need to accurately measure the angle and the circumference. On small cones, this is usually not a big deal. On large cones, it's difficult to get accurate results by using the angle to determine the pattern.

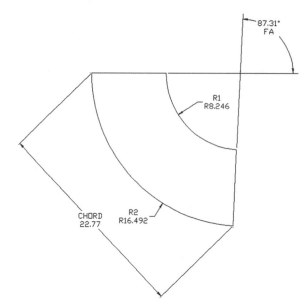

Figure 11-73: Key elements for frustum development.

The problem is the arc length defined by the angle becomes the circumference for the base diameter and the frustum diameter. If these are not the correct length, the cone diameters will be wrong. When measuring a large arc with an angular measuring tool, it's easy to miss by a quarter or half degree, affecting the finish diameter. It's always easier to measure a straight line like the chord than an accurate angle. Therefore, we just use the chord dimension to locate the end points on the pattern.

The second benefit of this method as that as soon as you have the chord dimension, you know how large a piece of material will be required. Note: if the flat angle is larger than 180 degrees, the chord dimension can't be used for sizing your starting piece of material. Instead, double the large radius dimension blank length and add a little bit for trimming.

Working through the example in Figures 11-72 and 11-73:

Base Diameter = 8.00   Height = 8.00
Frustum Diameter = 4.00

SH = Slant Height   FA = Flat Angle
CH = Chord

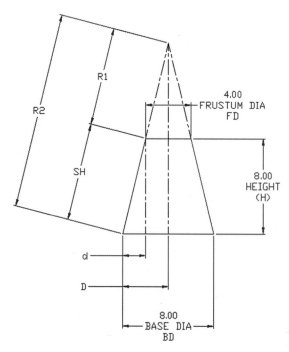

Figure 11-72: Frustum development.

$$SH = \sqrt{H^2 + d^2} \quad SH = \sqrt{8^2 + 2^2}$$
$$= 8.246$$

$$FA = \frac{d360}{SH} \quad FA = \frac{(2)360}{8.2462} = 87.3129°$$

$$R_2 = \frac{SHD}{d} \quad R_2 = \frac{(8.2462)4}{2} = 16.4924$$

$$R_1 = R_2 - SH \quad R_1 = 16.4924 - 8.2462$$
$$= 8.2462$$

$$CH = (Sin)\frac{1}{2}FA(2R) \quad CH = .6903(32.985)$$
$$= 22.7706$$

Figure 11-75: A holdback pin.

Try it out yourself with some little paper cones at your desk. You will be impressed with the accuracy of this method and the simplicity once you have done a few. I just noticed that the difference between the two radii in the example is exactly half. I didn't plan that when I set up the example cone; it must just be the math gods working their mojo.

When you form conical shapes in the power rolls, the ends of the cone blank need to travel through the rolls at different rates. Depending on the difference between the two diameters of the cone, this can be difficult to control while rolling. One little helpful trick is to use a stalling or holdback pin. The purpose of the pin is to retard the smaller end of the cone so it travels through the rolls at the same rate as the larger end. The block and pin shown in Figure 11-74 were fabricated to fit our small pyramid power rolls. The pin is spring

loaded so it can be pushed down into the block as the top roll is lowered.

Figure 11-75 illustrates using a holdback pin to stall the small end of a cone formed in the power rolls. What this holdback device looks like will depend on the configuration of the rolls you have. For initial pinch rolls, a simple length of angle iron, with a tongue the same thickness as your material and the Vee pointed into the pinch, will work for holding back the small end of a cone. The second desirable feature of this holdback device is it keeps the cone blank away from the housings of the rolls during rolling (Figure 11-76). Without the holdback device, the blank can drift and interfere with the machine.

Here is a funny little corner notch we use to fabricate cabinets with recessed doors (Figure 11-77). Surface-mounted doors always look lousy to me

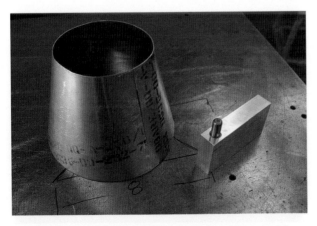

Figure 11-74: This block and pin fit small pyramid power rolls.

Figure 11-76: Keeping the cone blank away from the housings.

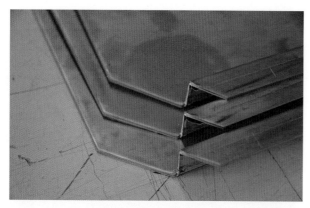

Figure 11-77: Corner notches for cabinets with recessed doors.

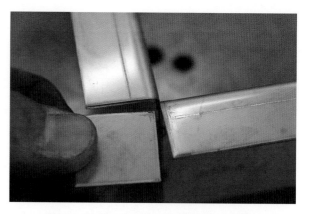

Figure 11-80: Turning the flange outside.

Figure 11-78: All the bends have been made.

Figure 11-81: The filler piece is larger than the space.

so we use this method to set the door into a recess. In Figure 11-78, all the bends have been made, creating a nice mitered corner with a door recess (Figure 11-79).

Here are ideas for rims and edges. Instead of doing a bunch of extra notching work on the sides of a box or tray-shaped part, it's sometimes easier to insert a separate filler piece. In Figure 11-80,

the flange is turned outside which means we either have to do a deep complicated notch or, in our case, add a simple filler piece.

Make sure the filler piece is slightly larger than the space (Figure 11-81). This allows your weld to go all the way to the outside edge and still leave a little material to sand back to make the flanges flush (Figure 11-82).

Figure 11-79: Creating a mitered corner with a door recess.

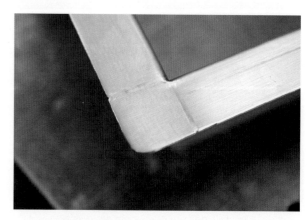

Figure 11-82: Making the flanges flush.

Figure 11-83: A tray with an extra flange

Figure 11-86: A small strip that follows the curve.

The tray in Figure 11-83 has an extra flange bent on it. This makes for a very stiff part, and a tray that is easy to pick up because of the extra bend. Filling in the corner is a little more complicated. Most folks just do a 45-degree filler piece, but I want to show a fancier rounded corner.

Figure 11-84 shows the start of the filler piece with notches. It's the flat piece that looks like a stealth bomber. After forming and fitting we end up with the part in Figure 11-85. This can now be welded in place.

Figure 11-87: The finished tray corner.

The last bit is a small strip that follows the curve and makes up the curved vertical part of the filler piece (Figure 11-86).

After a little sanding and detailing, we have a nice looking tray corner (Figure 11-87). Functionally, this is the same as the 45-degree variety, but I like the look of the fancier rounded corner.

Tanks and Baffles

Figure 11-84: The start of a filler piece with notches.

Many years ago my old sheet metal teacher taught me a way of installing baffles in tanks worth mentioning. Baffles are internal panels inside storage tanks and are used to control sloshing and shifting of the tank contents. Many tanks do not require baffles, but almost all tanks used in equipment that moves or is moved will be baffled. The requirements for baffles are that they are strong enough to resist the material inside the tank pushing against them, and that they slow but not block the passage of the mate-

Figure 11-85: After forming and fitting.

rial. For this reason, many baffles have holes or

Figure 11-88: A typical tank baffle.

Figure 11-90: The flange is partially consumed.

Figure 11-91: Using thin supports diagonally.

cutouts in them to allow the controlled passage and flow of the tank contents.

Figure 11-88 shows a typical tank baffle. The lower corners are clipped so material can pass back and forth between the two sides. In effect, a baffle creates two smaller tanks. The number of baffles required is related to the minimum size tank you would not bother to baffle.

Installing baffles in the tank body can be a little tricky. It is important for the baffle to fit the tank body properly so there are no exterior bulges or concavities, but still be easy to slip into place. Not too loose and not too tight is the right way for a baffle to fit.

The secret of this method is the short flange bent on three sides of the baffle (Figure 11-89). This makes installation of the baffle much easier. The flanges act like legs and keep the baffle in position for welding. They also make the size of the baffle adjustable by changing the bend angle of the flange. The third benefit of the flange comes during welding when the flange is partially consumed in

the weld as filler without damaging the tank body (Figure 11-90).

When you need to place a lid over an opening, a handy trick is to use thin supports diagonally (Figure 11-91). This trick still allows you to slide the lid around to get it tacked up without raising it too high. Use thin material so after you have a couple of good tacks on the lid, you can still slide the supports out (Figure 11-92).

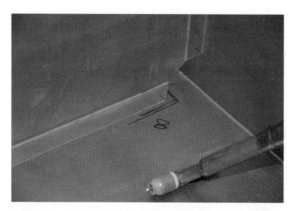

Figure 11-89: A short flange bent on three sides.

Figure 11-92: Sliding out the supports.

The Abrasion Department

Sanding, Grinding, and Abrading

A better title for this chapter might be "Finish Work." Many times, this important work is the last step for the machinist or fabricator before the job goes out the door or to the next department. This is the last chance we have to make a job look professional—or screw it up beyond repair. Bad finish work equals a bad job; it is hard to recover from it without a lot more finish work and that inevitable overworked look.

How many times have you worked so carefully for many hours, only to botch the last step like slipping with your de-burring tool or botching the part number marking, and really take an otherwise perfect

Figure 12-2: Going too far.

job spiraling down the drain? It might be something as dumb as dropping the part on the hard floor or filing a small burr off that is the cause of the damage.

As humans, we like to complete a task or accomplish something, but another little part of us cannot leave well enough alone when we really know better. I call this syndrome, "Just one more little thing." There is a critical juncture in every job where any further input by humans is really unnecessary and comes with an exponentially increasing risk of a screw up. Old timers have messed up enough so they recognize these forks in the road and heed the signs. They know when to get off the horse.

Figures 12-1 and 12-2 show before and after the point when somebody should have gotten off the

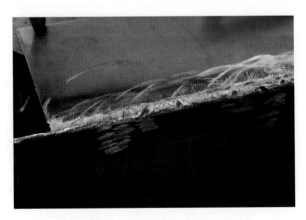

Figure 12-1: Quit while you're ahead!

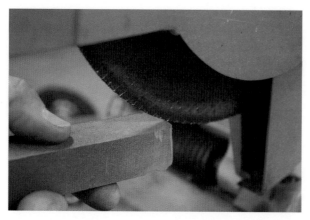

Figure 12-3: Using a stone to dress abrasive cutting discs.

Figure 12-5: Using a round grinding stone.

horse. . . . Don't try this at home, kids! No amount of additional grinding will help the part in Figure 12-1, as you can see in Figure 12-2.

Dress Cutting Discs. Dress abrasive cutting discs with a stone to maintain free cutting action (Figure 12-3). Using a stone is a trick that works to keep your cutting discs humming along. In general, these wheels break down quickly enough that fresh abrasives are exposed easily. The thicker the cutting disc, the better this technique works, especially when you have a broad cut in comparison to the width of the abrasive disc. Use this trick on chop saws or broad cuts in tough materials, or materials that tend to load the wheel, like stainless and aluminum. A quick touch-up with a dressing stone sharpens it in a flash. In Figure 12-4, I am using a boron nitride dressing stick called Norbide. This tough super-hard material makes a great cutting wheel dresser.

Tube Fitup. You can use a round grinding stone to make perfect joint fit ups in round tubing and pipe (Figure 12-5). These round grinding stones are available in several diameters. If I am fitting tubing that has an odd diameter, I would dress the round stone with a diamond in the lathe until it was a close match to the tube diameter. Good welding can only come after good fitup (Figure 12-6). This is one way to make the joints fit with a minimal gap. Use this method for those off-angle connections or the tube sizes that your notcher cannot do.

Graining Nails Made from Stainless Filler Rod. When you need to add a linear grain finish to stainless steel sheet, it can sometimes be a pain to clamp the sheet so you don't get funny runoff marks or stops and starts where the clamps were. One trick I have used is what I call "Graining Nails." I like to use a piece of plywood to grain flat

Figure 12-4: A boron nitride dressing stick (Norbide).

Figure 12-6: The importance of good fitup.

Figure 12-7: Graining nails.

Figure 12-9:
Grinder holders.

sheets manually. The plywood should be a little wider and a little longer than the sheet you are working on so you can clamp it securely. The graining nails look like small finishing nails and are made from snipped off pieces of stainless filler rod (Figure 12-7). We use stainless nails for stainless and steel nails for steel. Using alike materials prevents steel contamination of the stainless sheet from unwanted metal transfer. A few of these around the perimeter secures the sheet during graining without using clamps (Figure 12-8).

Cut Scotchbrite sheets down to smaller size. The full-size Scotchbrite hand pads are much too big for normal small blending and finishing jobs. Most of the time, you just need a little square to blend a little area. If you cut the Scotchbrite pads down into three finger-sized little pucks, you can fully use the abrasive and get many more square inches of blending out of each pad. Cut them with

a crummy pair of old tin snips. Never use your good snips to cut abrasives.

Grinder holders (Figure 12-9). You can tell if somebody knows how to grind by the way they lay their grinder down. The goal in good grinding is to keep the disc as flat as possible for the life of the disc. A warped disc makes for crappy grinding and jello arms at the end of the day. A better way to hang up a grinder with a sanding disc is to make a holder that protects the abrasive disc from damage (Figure 12-10).

Figure 12-11 shows the wrong way to lay a grinder or sander down. Most grinders have those nifty little black plastic bumpers on the back side. Why not use them? If you lay your grinder down like this for any length of time, the disc will warp and it will show up in your finish work. If you grind

Figure 12-8: Securing the sheet during graining.

Figure 12-10: Protecting the abrasive disc from damage.

Figure 12-11: The wrong way to lay down a grinder or sander.

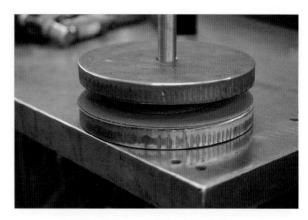

Figure 12-13: A shop-made flattener.

stainless steel, laying your abrasives down like this on a steel table will contaminate the abrasive and transfer the contamination to the stainless steel.

Sanding disc flattener. It's a real pain when your sanding discs are warped from improper storage or high humidity. Everything is fine until you open the package (Figure 12-12). Warped discs only cut on the high points, so it's important that your discs are pretty flat if you want to do decent work and get the most from your premium abrasive dollars. The simple shop-made flattener in Figure 12-13 keeps a little pressure on the discs as the humidity changes.

Another trick to extend the life of your sanding discs is to rotate them after you have used them for a while. The trick is to rotate them in relation to the backing pad. As you can see in Figure 12-14, this disc has not been cutting evenly all the way around. The spot I am pointing at has seen almost no work.

When you rotate it slightly in relation to the backing pad, you can sometimes move the low point of the disc to an area on the backing pad that is higher.

Wrap abrasive around a stiff flat bar for better cutting action (Figure 12-15). The hard backing of a metal bar converts more of your elbow grease into removed metal. Thin flat-bars sneak into gaps

Figure 12-14: Rotating the backing pad.

Figure 12-12: Warped discs.

Figure 12-15: Wrapping abrasive around a flat bar.

Figure 12-16: Using thin flat bars.

Figure 12-19: Adding a bit of oil or WD-40.

and openings where regular sanding pads won't fit (Figure 12-16). It might be helpful to think of this idea like a fine grit file. The action is very similar. You can use a small clamp to secure the abrasive from moving.

Tape abrasive paper to the surface plate to keep the wrinkles out (Figure 12-17). I hate it when

Figure 12-17: Keeping wrinkles out.

Figure 12-18: Changing the direction of sanding.

you get a nice fresh sheet of silicon carbide paper out to do a little flat lapping and the first thing that happens is you crinkle the paper and ruin it for lapping. This trick allows you to think about the lapping job instead of concentrating on holding the paper steady.

Change the direction of sanding to give a balanced crosshatch pattern (Figure 12-18). I call this fake surface grinding. It's not meant to deceive anybody, instead just to mimic a Blanchard ground surface. In some cases, it looks better than the standard figure eight pattern normally used. It also helps you see the scratch lines left from the previous grit abrasive paper.

Add a little light machine oil or WD-40 to make an abrasive act like a finer grit (Figure 12-19). This trick reduces the effective grit by approximately half, allowing you to do your finish with the same abrasive grit (Figure 12-20). This also works well with

Figure 12-20: Reducing effective grit by half.

Figure 12-21: Internal ball hones.

Figure 12-23: De-burring cross-drilled holes in a bore.

Figure 12-22: Using hones with a drill.

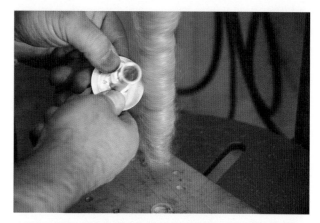

Figure 12-24: Wrapping bronze wool around threaded rod.

the ball holes mentioned elsewhere. The kerosene in the WD-40 keeps the abrasive clean and cutting.

Change direction 90 degrees each time you change abrasive grits. This makes it easy to see the scratch lines from the previous grit. There is nothing worse than trying to get 100 grit lines out with 600 grit paper. I think the numbers mean the number of hours to remove the scratches . . . !

Ball Hones. Internal ball hones have saved my bacon quite a few times (Figure 12-21). These cheap and easy-to-use hones have a dozen uses around any shop. They work equally well on soft and hard materials using a common cordless drill (Figure 12-22). I have even used these in a pinch to resize a commercial ball bearing bore to fit a slightly oversized shaft. They can de-burr those nasty cross-drilled holes and steps in a bore (Figure 12-23) as well as improve the surface finish an easy ten points.

DeBurr in the Drill Press. Here is an interesting de-burring trick. Take steel wool or, in the case of this example, bronze wool and wrap it around a piece of threaded rod in the drill press (Figure 12-24). This mild cutting action is just the ticket for many delicate de-burring jobs. The threaded rod has the right amount of tooth to retain the wool and keep it from spinning on the rod (Figure 12-25). The cutting action is similar to a non-woven abrasive, but with a greater polishing ability.

Stainless Polishing. Figure 12-26 shows one of the best kits I have ever seen for producing fine finishes on stainless steel. The kit consists of all the abrasives needed to go from weld grinding the entire way to a mirror finish. They are all correct for the incremental finishing needed to produce a high polish on stainless and many other metals. Walter Abrasives has collected all these grits into a great little kit called "Quickstep."

Figure 12-25: Cutting action similar to non-woven abrasive.

Figure 12-27: A stainless flat bar.

Figure 12-28: Starting with the first flap disc.

Our test volunteer is a piece of stock mill finish stainless flat bar (Figure 12-27.) You can see the rough mill finish and even a boot print where somebody walked on it!

Starting with the first flap disc in the set (Figure 12-28), each abrasive in the set is just the right amount finer for each finishing step. Be sure to change sanding direction as you work your way up into the finer grits. If you do not remove all the scratch marks from each grit, you will have to go back and remove them with a coarser grade.

Now it's hard to take a picture of a highly reflective surface, but I think the one in Figure 12-29 does the results justice. You can see the reflection of the screw of the Bessey clamp in the highly polished surface. One thing I have found important while doing

Figure 12-26: A kit for producing fine finishes.

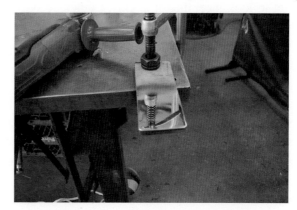

Figure 12-29: A highly reflective surface.

Figure 12-30: Squaring the table of the belt sander.

Figure 12-31: Replaceable front edges.

Figure 12-32: Reducing the gap.

high finishes is to limit the heat in the part. Cool it off once in a while to keep from melting or damaging the finer abrasives in the kit. I spray water on the part and blow it off with an air hose to cool it between grits.

Belt Sander Table. Square the table of the belt sander before you start (Figure 12-30). The table tends to droop downward during use if not checked. Therefore, if you have a finish squaring job that you want to look nice, then it is only a quick second to check and adjust for perfect results. On this same note, the belt platens will need replacement or re-surfacing every few years to maintain a decent flat backing for nice detailing work.

In Figure 12-31, we have added replaceable front edges to the belt sander tables. This allows us to keep the gap between the belt and the table close and not have to replace the table casting every year.

When you have to detail very small parts, a handy trick is to use a piece of flat bar to reduce

the gap between the belt or disc and the worktable (Figures 12-32). This prevents your tiny part or, worse, your finger from getting sucked into the gap and lost forever (Figure 12-33).

If you have a bunch of parts that get sanded at a particular angle, you can mark witness lines on the

Figure 12-33: Protecting the tiny part . . . and your finger.

Figure 12-34: Marking witness lines.

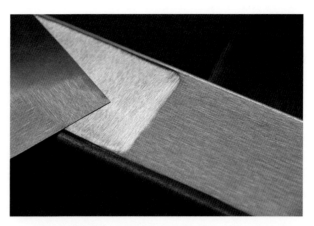

Figure 12-36: Changing the graining direction.

Figure 12-35: Using a thin masking plate.

Figure 12-37: Marking the materials.

table of the belt sander to guide your roughing work (Figure 12-34). This visual cue keeps you from drifting too far from your intended angle. You can clamp a guide to the table, but that destroys one spot on the belt.

Masking Plate. Use a thin masking plate when you have intersecting linear finish lines. This produces a crisp demarcation line between the two intersecting finishes. Be sure to use a plate of the same material to avoid contamination. In Figure 12-35, the masking plate is .010 stainless shim stock.

The long length was grained first, then the mask was used when the graining direction was changed (Figure 12-36). It's easier if you have something to grain with a little length to it, but this demonstration gets the idea across.

Mark Abrasives. Mark the materials you use your abrasives on and segregate them from the other discs (Figure 12-37). I have seen some beautiful jobs

ruined by a contaminated sanding disc or even a dirty wire brush. When in doubt, use a new disc. One of the guiltiest parties in the shop with regard to contamination is the dirty little wire brush on the bench grinder or buffer (Figure 12-38). It spreads its corrosive disease to everything it touches. You may

Figure 12-38: A contaminated wire brush.

have the best intentions, but this wheel is almost always contaminated with steel, rust, and grease in the average shop.

Laps and Lapping. Figure 12-39 shows a few examples of laps and lapping. These are all cylindrical laps for OD and ID work. They are quite simple to make in the shop and produce some fantastic results. These particular brass laps were used to lap a bearing bore (far right) and the OD of a precision gage (center). The results are controllable down to micro-inches if necessary. In this case, the lapping compound was diamond paste. The laps themselves need to be accurately made, but with a little effort results that cannot be attained any other way are possible. All the laps shown in Figure 12-39 are used under power at slow speeds, approximately 100–300 RPM. and stroked axially over the length to be lapped to produce a crosshatch pattern of approx 30–45 degrees.

Dress your surface grinder wheels at a slight angle for peripheral roughing work (Figure 12-40). The edge will break down faster, exposing sharp grains for faster stock removal. Re-dress for the fine finish work. For roughing, use an aggressive depth of cut and step over as much as the wheel will tolerate without complaint. The abrasive grains need to break down to continue the cutting action. If they are babied and allowed to glaze over, all your effort just turns into heat. Abrasives should be thought of as little cutting tools, much like a lathe

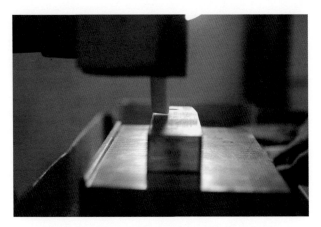

Figure 12-40: Dressing surface grinder wheels.

or milling machine cutting tools. Properly used, the chips look similar under high magnification.

A similar roughing trick works for side wheel grinding, which on a good day is a bit of a pain. Side wheel grinding is not an efficient material removal method, but is necessary some times because of setup issues or part geometry (Figure 12-41). Dress the side of your wheel as you would normally. Stop the grinder and loosen the nut. Rotate the wheel slightly in relation to the spindle. Tighten it all back up and go ahead and do your side wheel roughing. What happens is the wheel will run out a little from where it was dressed and only cut on one part of the periphery. This small edge breaks down quickly and continues to cut with the need for frequent redressing during your roughing. Dress again and don't rotate the wheel for your finish work. This trick can give you more

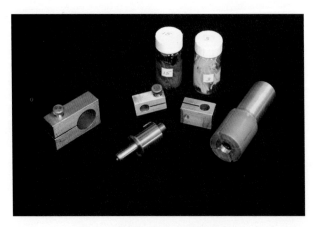

Figure 12-39: Laps and lapping.

Figure 12-41: Side wheel grinding.

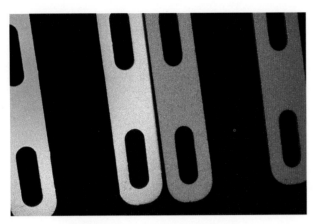
Figure 12-42: Glass bead blasting.

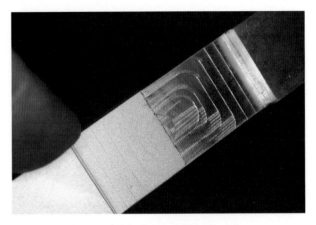
Figure 12-44: Removing tool marks.

grinding time between dressings, which makes a difficult grinding job go faster.

Bead Blaster Finishes. Glass bead blasting can de-burr and produce finishes that are hard to produce any other way. Anything from a fine satin finish all the way to a rough surface to add grip or traction to a part can be produced with a few different grits and by varying the blasting pressures used (Figure 12-42). Tool marks can be blended or removed with fine mesh size glass beads (Figures 12-43 and 12-44). Beware of blasting large sheets and surfaces. The bead blaster is like hammering with a billion little ball pein hammers; it will move material if you're not careful. The bead or sand blaster is another source of material cross contamination that is often overlooked. If you have been cleaning engine parts for your jeep, you probably need to replace the media if you're going to blast some medical parts.

Ever had a hard time holding onto small parts in the bead blaster? It's really a pain if you drop one in the bottom of the blaster and have to fish it out. This little tea strainer trick keeps the part under the blast stream and captured (Figure 12-45). Be sure the mesh opening is larger than your beads or it will take a really long time to do the job. I have shown this opened up so you can see inside. A spring closes the two halves together when in use.

Patterns and logos can be etched or imprinted on a surface by using a simple mask and abrasive blasting. The mask in Figure 12-46 was laser cut for this specific purpose to produce a pattern on glass sheet. In this example, I did a piece of polycarbonate sheet and left the factory masking paper on the plastic sheet while I blasted it (Figure 12-47).

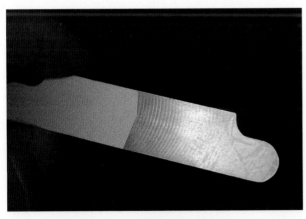
Figure 12-43: Blending tool marks.

Figure 12-45: A tea strainer trick.

Figure 12-46: A laser cut mask.

Figure 12-48: Using carpet samples.

Carpet Protectors. Now you have a use for all those cheap, nicely-hemmed carpet samples they sell at the flooring stores (Figure 12-48). They make great throw-away portable part protectors and backup surfaces for DA detailing. When they get loaded with dust, toss them out and start again.

Surgical DeBurring. For fine de-burring, try a surgical scalpel blade (Figure 12-49). The sterile stainless blades seem to be considerably sharper than the plain Jane carbon steel models. If you don't like it for de-burring, you can always do a little amateur shop surgery with it!

Ceramic DeBurr Knife. The ceramic-bladed scraper seen in Figures 12-50 and 12-51 is superior for gummy grabby plastics like UHMW and ABS. These materials can be a challenge when trying to do a clean, crisp de-burring job. The ceramic has just the right amount of edge sharpness to peel, without gouging softer materials.

Another trick is to put soft plastics like this in the freezer if you have the time. This makes the burrs more frangible and easier to snap off cleanly.

Vibratory Tumblers. Vibratory tumblers are a double-edged sword with the potential to decrease your hand de-burring time by a huge amount or destroy an otherwise simple job in a matter of

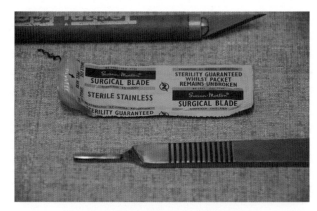

Figure 12-49: A surgical scalpel blade.

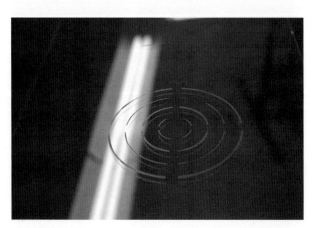

Figure 12-47: Imprinting a polycarbonate sheet.

Figure 12-50: A ceramic-bladed scraper.

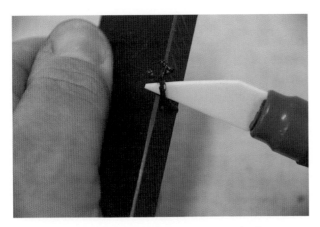

Figure 12-51: Working with gummy plastics.

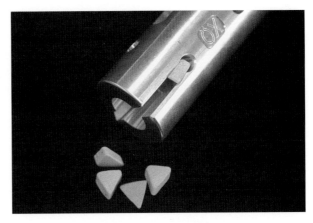

Figure 12-54: Clogging.

minutes (Figure 12-52). The secret is all in the media and the part geometry. The deadly booby traps are media wedging and plugging (Figures 12-53 and 12-54), and material cross-contamination, not to mention just plain losing small parts. You can lose huge amounts of time just trying to find small parts if you're not set up properly. Run sample extra parts first before you commit a large batch of critical parts. Beware of overloading larger parts that can clank together causing serious damage. It's a real pain to remove ceramic media from blind tapped holes without messing something up. Always note a proper count when you drop a batch in. There's nothing like finding that one missing part two months later when you change media.

For really small parts, you can use a sub-container inside the main bowl (Figure 12-55). This prevents tiny gray parts from becoming hopelessly lost in a sea of tiny gray media. It's also a neat trick to allow you to switch media sizes and shapes without having to dump out 300 pounds of rocks for your one little part.

You can mask features you want to protect or stay sharp while the rest of the part goes through the vibratory tumbler de-burring cycle (Figure 12-56). Plastic or rubber caps and the mesh sleeve used to

Figure 12-52: Vibratory tumblers.

Figure 12-53: Media wedging.

Figure 12-55: Using a sub-container.

Figure 12-56: Masking features for protection.

Figure 12-57: Securing parts to a leash.

protect shafting work great for masking. You can also use small silicone plugs for masking holes you don't want media to enter.

You can also secure small parts to a leash of sorts to make them easier to retrieve from the tumbler (Figure 12-57). I use stainless tie wire and a large "flag" that I can spot easily when I open the lid. This cuts down on the inevitable search and count associated with media tumblers without parts discharge devices.

The Good, the Bad, and the Ugly

Little grinders are for little work. If you have any serious material to remove, do yourself a favor and get a large sander or grinder. Seven-inch seems to be the optimal size for normal-sized people.

Nine-inch diameter wheels have an annoying gyroscopic effect that tends to fatigue the operator more quickly without much metal removal advantage. Seven-inch abrasives also seem to be more readily stocked and available in a wide array of abrasive materials and grits.

I am a firm believer that you get what you pay for in abrasives. Cheap abrasives are just that—cheap. The longer an abrasive disc lasts before it glazes and just produces heat is the true measure of performance. For sheer metal removal and fine control, I would put a coarse sanding disc up against any hard type disc for metal removal rate any day (Figure 12-58). The only exception to that would be in an application where the amount of abrasives used per hour outweighed the labor costs for changing discs. In other words, sanding discs cut faster, but don't last as long as hard discs.

Hard discs can do some things sanding discs cannot do, like cut a bolt off flush with the floor or any application that needs the front edge of the abrasive wheel to remove metal (Figure 12-59). The main advantage of the hard discs is for slot or recess grinding and longevity.

Another application where hard discs are superior to sanding discs is in a situation when you have hot-rolled mill scale to remove. The hard disc fractures this tough-to-grind material (Figure 12-60) whereas a sanding disc just polishes

Figure 12-58: A coarse sanding disc.

Figure 12-59: Advantages of hard discs.

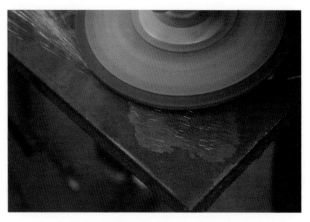

Figure 12-61: Polishing tough-to-grind material.

itself into oblivion (Figure 12-61). Hard discs have their place in every shop, don't get me wrong. They last longer, but don't cut nearly as fast. At today's labor rates, which would you prefer to pay for?

Most people wait way too long to change discs. The telltale signs are excessive heat and a polished surface. The sanded surface should have a matt or satin finish to it and appear flat and non reflective when the abrasive is cutting properly. If you examine the disc, you can see the polished surface of the glazed abrasive surface grains. Sometimes you can dress these out with a dressing stick to get a little more life out of the disc. I tear the disc when I throw it in the trash so the cheapskate dumpster divers don't pull them out and waste time with discs that are worn out.

Radius Grinding

A technique I call roll grinding is used to finish corners to match bend radii. Long weld seams that need to be rounded can be handled using a rolling technique with the disc sander. It takes a little practice, but once you have it mastered the results look just like a formed corner. In one shop I worked in, we did miles of seams like this. Good weld technique can minimize the amount of metal removal for this fine finishing technique.

The weld should first be decked off flush against the two intersecting right angle surfaces (Figure 12-62). For bumpy welds, I sometimes make a few light faceting passes to even everything out before the rolling (Figure 12-63). If you're right handed, stand with the weld to your

Figure 12-60: Fracturing tough-to-grind material.

Figure 12-62: Decking off the weld.

Figure 12-63: Light faceting passes.

Figure 12-65: Rolling action with the left hand.

right. Your right hand is the balance hand that supports the weight of the grinder and your left hand does the precision rolling motion. The wheel or disc should be kept as flat as possible on the weld seam. Very little pressure is used to do the actual rounding. Think of your right hand as the pivot holding a round object like a motorcycle grip or a piece of round handrail. As you sand, step slowly backward along the seam.

Your rolling passes should overlap by three-quarters of the disc diameter or more (Figure 12-64). Remember: this is not an aggressive action, but a fine finishing action. The rolling action with the left hand should go flat to flat through 90 degrees or just a little shy (Figure 12-65).

It's pretty hard to take a picture of the end results. If you did everything right, it should look almost exactly like a formed bend produced on a brake (Figure 12-66).

Paint Fill Myth. The paint will fill in the scratches. This myth continues to stay alive. The best way to describe why it doesn't work is look at some "Snow Capped Peaks." Just because a little snow is on the mountain, doesn't mean you can't see the mountain (Figure 12-67). Parts show grinding scratches after a light coat of flat paint. Paint is

Figure 12-66: The end result.

Figure 12-64: Rolling passes.

Figure 12-67: Even paint shows scratches.

a coating that has some thickness, but generally not enough to hide bad grinding. Just ask any auto body shop that does painting how smooth the surface should be prior to painting. The underlying surface needs to be smooth and scratch free. Because most painted surfaces have some gloss to them, they tend to highlight any scratches that were not removed.

Figure 12-68 is the same surface shown previously, but I used a finer grit abrasive and a final skim with a non-woven abrasive like Scotchbrite to remove most of the offending deep scratches.

The finish of the finishing chapter has arrived!

Figure 12-68: Removing the scratches.

The Junk Drawer

Miscellaneous Tricks Without a Home

This chapter is like that special odds and end drawer in your toolbox. You know, the one with all the weird stuff in it for which you can't seem to find the perfect storage space. Almost everybody has at least one. One of the more famous ones is the drawer in everybody's kitchen that has all the diabolical food preparation tools that always seem to interlock in ways that make them difficult to remove from the drawer. It's also the drawer where the tool you most want is *always* on the bottom.

When I was a kid my dad had a workshop in the basement. I spent a fair amount of time snooping around because that's what kids do when left to their own devices. There was a special drawer my dad called the top drawer. It was the top-most drawer in a cabinet under the workbench right next to the DC bench grinder my dad got from a navy ship. It was a magical drawer. It had all kinds of strange and interesting things in it. Ever seen a cork boring set? How about the tool to sharpen the cork boring set? Master links, steel balls, three pounds of hex wrenches left over from mounted bearing kits, springs, a veritable mechanical cornucopia. If I was on a deserted island this would be the drawer I would want with me.

This chapter is kind of like that top drawer. A mix of experiences and techniques that don't fit very well anywhere else, but just like the top drawer they are worth keeping.

By the way, machinists and other metalworkers all have drawers like this. Just for fun, here is a list of some things that I would bet money you would find in most machinists' toolboxes or top drawers:

- Steel balls from ball bearings. Machinists can't throw these away, ever.
- A little box of carbide inserts that don't fit any of the tool holders in their box or in the shop.
- Three-to-five insert torx wrenches all the same size. These are way too nice to just throw away.
- At least one broken 6-inch scale.
- A pound or more of dowel pins in assorted sizes and lengths.
- Several pieces of brass rod from 1/4 or 3/8 in diameter, about 3 inches long. Mushroomed ends optional.

If you do any welding work, add the following to the list,

- At least one tape measure with the end broken off.
- A ball pein hammer with tape around the handle up near the head.
- A pair of vise grips missing the spring.

Shop Heat Treating

When wrapping parts in stainless wrap for heat treating, add a pinch of fine cast iron shavings to the envelope (Figure 13-1). These burn up during the heat treat and consume the detrimental oxygen inside the wrapping. The small particles of cast iron heat up and burn long before the heavier parts have a chance to scale from the oxygen (Figure 13-2).

Another shop heat treating tidbit is to blow some Argon from the TIG welding torch into the envelope to help exclude oxygen and improve the atmosphere inside the envelope (Figure 13-3).

Tiny Parts

Sweep the floor before you work with really small parts and assemblies. Those tiny parts resemble chips and debris pretty closely (Figure 13-4). I'm getting too old to crawl around on the floor

Figure 13-3: Another heat treating tidbit.

so I had to get a little smarter. The precious part stands out in stark relief on a cleanly swept floor (Figure 13-5).

Put something in the sink drain before you wash those tiny little parts you just spent three days making (Figure 13-6). I really hate it when I lose a part

Figure 13-1: Fine cast iron shavings.

Figure 13-4: Working with tiny parts.

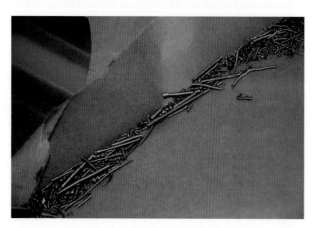

Figure 13-2: Working with shavings.

Figure 13-5: Advantages of a cleanly swept floor.

Figure 13-6: Blocking the sink drain.

to plain old stupidity. This beats explaining why you're taking the drain trap apart.

If you have a really diabolical mechanical assembly that is a jack in the box of preloaded balls and springs, do yourself a favor and put the thing inside a clear plastic bag to take it apart (Figure 13-7). Slip

your hands inside the bag or do the work from the outside if possible. If you absolutely cannot lose any parts, this is the way to keep them at least contained to a small area.

Safety Wire

I love annealed stainless safety wire (Figure 13-8). This stuff is so consistent and pliable, it's like metal taffy. I use it to secure small parts for tricky welding or silver soldering jobs (Figure 13-9).

Stainless safety wire can stand up to the heat of a welding torch without breaking or melting away while still holding your assembly in position (Figure 13-10).

Studs and Standoffs

Many times machinists are called upon to cut small lengths of threaded rod to use as studs or connectors in an assembly. Instead of fussing

Figure 13-7: A mechanical assembly with many parts.

Figure 13-9: Securing small parts.

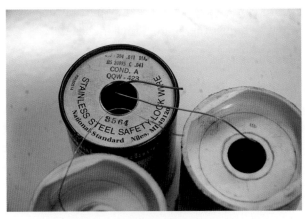

Figure 13-8: Annealed stainless safety wire.

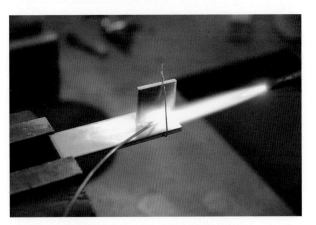

Figure 13-10: Stainless safety wire.

Figure 13-11: Long set screws.

Figure 13-12: When needing a solid standoff with a male thread.

Figure 13-13: Moving the machine.

around with cutting and the inevitable de-burring of all-thread, we buy several popular sizes of long set screws (Figure 13-11). They have the nifty added feature of a small hex driver in the end so you can hold or tighten them. One of our favorite tricks using set screws is in a situation when you need a solid standoff with a male thread (Figure 13-12). We fabricate the standoff with two tapped holes, which is pretty much cream cheese for any shop in any material. Instead of single pointing a male thread, we just install a long set screw into the female thread and—presto!—instant male thread.

Ideas for the Shop Floor

Sometimes it's easier to move the machine to the work instead of trying to maneuver awkward material to the machine (Figure 13-13). We regularly spin the milling machines around so long stock can

go out the door. We sometimes drag the floor-mounted drill press outside to work on long stock that would be a real hassle to handle inside.

Before you try to position a tricky part with one hand, be sure to have all your clamps preset close to the correct openings so you can work them with your remaining free hand. There's nothing worse than having to set down a heavy part because you forgot to adjust the clamps four inches.

Tube Forming

When in doubt, try it out. Here are a couple of examples of tube end forming that we initially thought were going to be very difficult, but turned out to be very easy. The hexagon was just that—a male mandrel was pressed into the tube with lubricant and then withdrawn (Figure 13-14). The flats were squeezed in a smooth-jawed vise; it fit the hex with a play free slip. The expanded section of this

Figure 13-14: Tube end forming.

Figure 13-16: Stretching and straightening small diameter tubing.

Figure 13-15: Expanding a Hastelloy tube.

Figure 13-17: Tube fabrications.

Hastelloy tube was done in small steps with 5C expanding collets (Figure 13-15). Originally the customer wanted the tubes machined from solid. We tried a quick idea with the tubing and made a better part in the end with a lot less frustration. Sometimes all it takes is a willingness to try a wacky idea. Many times with a little curiosity and effort you end up with a new skill and the confidence to use it.

The pivoting fixture was used in the press brake to stretch and straighten small diameter tubing (Figure 13-16). The job was to resize a tube to make a slip fit on the ID of another tube. The matching tube diameter was not commercially available, so we took the next closest size and made a drawing fixture to reduce its diameter. The stretching had the added benefit of making the tube as straight as an arrow. We were able to control the diameter within a couple of thousandths of an inch without trouble.

The way this works is one end of the tubing was anchored to a fixed block off to the left. The opposite end was mounted to this moving block. We used swagelock ferrule fittings to lock the tubing into the fixture. The press brake was fitted with a flattening punch which came down on the roller in the top of the picture. By setting different depths, we were able to control the stretch of the tube very closely.

Tube Coiling

Figure 13-17 shows a couple of interesting tube fabrications. They're interesting because we do not have any machinery for coiling tubing. The material is soft copper refrigeration tubing and were done with one piece of tubing. The first step was to uncoil the raw material and straighten it for our forming operation. This was 1/2-inch outside diameter, so it took a little force to straighten it. We ended up using the forklift to assist in the

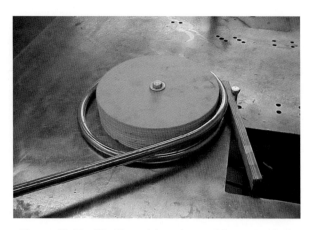

Figure 13-18: Working with a plywood form mandrel.

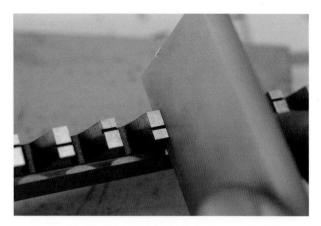

Figure 13-20: Stoning the edges of a new broach.

straightening. We secured one end of the tubing to one of our heavy welding tables; the other end we attached to the mast of the forklift. Using the hydraulic tilt feature, we pulled the tubing perfectly straight without any kinks or bumps. We then hand wound the tubing on a plywood mandrel similar to the one in Figure 13-18. The size of the form mandrel was determined through testing so with the tubing spring back we ended up with the correct size coil.

Figure 13-18 shows the same tube coiling method, but a different project. The plywood form mandrel is just stacked up as high as you need the coil. The plywood mold is bolted to the welding table and a stop wedge inserted to hold the free end.

As in many cases with cutting tools, they can actually be too sharp to cut properly (Figure 13-19). The shape of the broach teeth naturally causes the

lower teeth to bite in and tilt the tool. Once the broach tilts, it is cutting the keyway too deep—typically at the bottom where you can't see what's going on. You can carefully stone the edges of a new broach to keep it from digging in (Figure 13-20). Before you folks jump down my throat and tell me I'm a dimwit, I learned this trick from a guy at Dumont Broach. We were having problems broaching some stainless sleeves, so I called to discuss the problem. It was a very enlightening discussion and solved our problem. As a habit, I back off the pressure several times during the broaching operation and check the alignment of the broach to the axis of the work.

Metal Tape Uses

You can use soft aluminum tape to help prevent breakout in delicate materials (Figure 13-21). The tape is just rigid enough to support those delicate

Figure 13-19: Cutting tools that are too sharp.

Figure 13-21: Preventing breakout.

Figure 13-22: Using soft aluminum tape.

Figure 13-24: Machining M2 drill blanks.

corners. This trick, combined with modifying your tool path so the cut path is toward the interior of the part, can get you out of a sometimes difficult operation (Figure 13-22).

Figure 13-23 shows another great use for soft aluminum tape. This tape will hold up under forming pressure to protect the finish on your materials. It can also be used to prevent iron transfer from the press brake die to a critical work piece.

Machine after Heat Treat

Try machining some parts after heat treating. 17-4 and 15-5 stainless steel are great candidates. In fact, some of these alloys actually increase in machinability rating with a mid-range heat treat. We have also been able to machine M2 drill blanks directly into cutting tools in a pinch (Figures 13-24 and 13-25). This beats a bunch of tricky setups on the surface grinder any day.

Deep Hole Drilling

For small-diameter, deep-hole drilling, use drills with a flatter-included angle 135 or more (Figure 13-26). They drill straighter than steeper angles. Short pecks 1/2–1 diameter, with a full retract, are the norm when drilling deep 10+ diameter depth holes (Figure 13-27).

Gasket Cutting

Use the end grain of a block of wood for cutting your gaskets and other soft materials with a hand

Figure 13-25: An improvement over surface grinders.

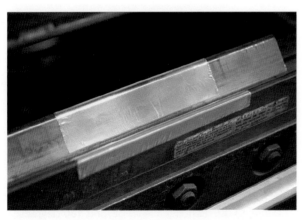

Figure 13-23: Another use for soft aluminum tape.

Figure 13-26: Drills with a flatter-included angle.

Figure 13-27: When drilling deep diameter holes.

Figure 13-28: Cutting gaskets with a hand punch.

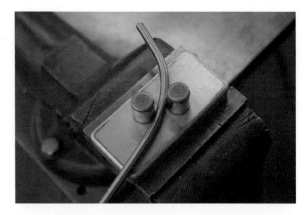

Figure 13-30: Making hooks and loops.

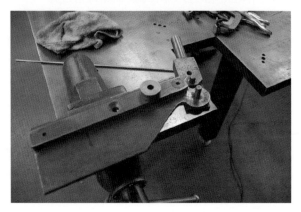

Figure 13-31: A tool for bending steel rod.

punch (Figure 13-28). The end grain of the wood exposes the tubular structure of the wood and allows the punch to penetrate cleanly through your material and into the block.

Rod Bending

Here is a little tool I made for straightening and bending small rods (Figure 13-29). It clamps in the

bench vise and has a step so the vise can grab it. The pins are plain old dowel pins. This tool is handy if you make hooks and loops out small-diameter rod and wire (Figure 13-30).

I built the tool in Figure 13-31 many years ago at a place where we had to bend miles of 1/4-diameter steel rod. These bent rods were piped all over the inside of the device; the electricians bundled their

Figure 13-29: A tool for straightening and bending small rods.

Figure 13-32: Bending a loop beyond 180 degrees.

Figure 13-33: Another homemade bending tool.

Figure 13-35: Extracting the bent tube.

Figure 13-34: The arm pivots and carries a roller.

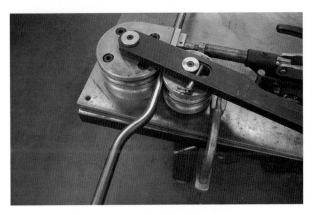

Figure 13-36: Making deep offsets.

wires to the looms to make a neat job of the wiring. The bender clamps in the vise and can bend a loop beyond 180 degrees (Figure 13-32).

More Tube Bending

Figure 13-33 features another homemade bending tool. This one was first built to bend the tubing nerf bars on racing go-karts. It bends 1/2 and 3/4 diameter tubing. As are most things I build, it was made from scrounged leftovers from paying jobs. My version of industrial recycling.

The arm pivots on a bearing and carries a roller with rounded grooves in it to follow the bend (Figure 13-34). The cross pin releases the following roller so you can extract the bent tube from the machine more easily after multiple bends (Figure 13-35).

With the roller follower close, fairly deep offsets are possible (Figure 13-36). When I want to use the bender, I just clamp it to the welding table. I never bothered to bolt it down anywhere.

Closing Thoughts

So, it looks like we have come to the end of the line—at least for this volume. Boy, am I happy! One, for actually finishing the book. And another, because it turned out to be a bigger chunk than I expected, and I need a vacation!! Seriously, working on *Sink or Swim* helped rekindle my deeper passion for metalworking. Without that excitement about the trade, I could have never finished it.

This was a pretty big project all said and done—in particular, the management of all the pictures needed to bring the book together. I shot something like 15,000 pictures in all, of which only a small portion actually ended up in the book. If you ever need a picture of anything in a metalworking shop, let me know; I might have one.

In the process of writing *Sink or Swim*, I learned many new things about myself and my trade. I kind of became a metalworking rancher. Let me explain.

A typical rancher has to be skilled in many different careers besides cows and crops. An average rancher ends up becoming proficient at operating and repairing heavy equipment to maintain fields and roads as well as maintain dozens of fossil fuel burning contraptions like pumps and chainsaws. Any decent rancher can negotiate a business contract, buy a hay baler, and not lose his shirt, as well as fell a tree or brand a steer.

What I am getting at is I had to learn about a lot of things I didn't know much about to write this book. Working with the editor and publisher improved my writing and empathy for the reading audience. It also improved my ability to take skills that I learned by doing and explain them in words and writing. Taking the thousands of pictures for this book taught me ways to get better pictures with less effort. My goal has always been to capture as much knowledge about metalworking and get it down on paper so other people can make use of it and have some fun in the process. I hope this book has achieved that to some level.

So if you think you might want to write a book about your skills and life experiences, here is a short list of booby traps I fell into that you might want to avoid.

- A little bit of work on your book frequently is better that a bunch not very often.
- When you think you have taken enough pictures of something you are describing, take three more.
- When you think of something you want to include in your book, quick, quick, write a note so you will remember. I adopted the habit of carrying a small notebook in my wallet for this exact purpose.

Photography

Most of the pictures in this book were shot with a Nikon D70 digital camera. A few of the pictures that appear were older photos I had taken with an

299

older lower resolution camera; many of them I couldn't re-shoot because of the subject matter. I used only two lenses for all the pictures—either a Nikon 18-70mm or a Sigma 1:1 macro.

Photographic lighting was difficult. I used extra lighting only in special occasions. Moving around the shop from one bay to another made setting up fancy lighting quite a bit of extra work, so most were shot with ambient shop light. I did use a tripod when in the shop. Not for every shot, but as many as I could. This is especially important with the macro work.

Organizing the photos would have been nearly impossible without the help of a photo organizing program. There are many out there, but I used Picasa by Google. It was free and had most of the features I needed to organize all the artwork in the book.

I truly hope you enjoyed this book. I certainly had fun and learned a lot writing it. If you have any questions or comments, or would just like to share a great shop story, you can contact me via email at:

- Positive feedback, movie offers, shop tours, etc. sinkorswimbook@gmail.com
- Negative comments, whining, bellyaching, and redundant questions, etc. deletemeorsink@gmail.not

Remember: keep learning everything you can about your trade. It's your best protection from early fossilization. Good luck and may your chips be tan and your tools always sharp!

Recommended Reading List

Title	Author or Publisher
McMaster Carr Catalog	McMaster Carr Supply
Hard to get but, yes, read it.	
Machinery' Handbook	Industrial Press
If you don't have a copy of this book, you aren't a machinist.	
Jorgenson Stock Guide	EMJ Metals
Now called EMJ on the web, you can sometimes weedle one of the stock guides from a salesman. Otherwise, look for an old copy.	
Design of Welded Structures	Omer Blodgett
Read anything written by Mr. Blodgett.	
Design Ideas for Weldments	Lincoln Arc Welding Foundation
Machine Shop Trade Secrets	James Harvey
Metals Handbook V3 Machining	American Society of Metals
Metals Handbook V4 Forming	American Society of Metals
CNC Programming	Peter Smid
Practical Ideas for Metalworking Operations	American Machinist
Morse Tools Machinist Handy Book	Morse Tool Company
Tool Engineers Handbook	McGraw-Hill
Details for Product Design	Greenwood—McGraw-Hill

Illustrated Sourcebook of Mechanical Components	Parmley
Foundations of Mechanical Accuracy	Moore Tool Co.
Holes Contours and Surfaces	Moore Tool Co.
Mechanical Drawing	French
Engineer to Win	Carroll Smith
Prepare to Win	Carroll Smith
Mechanisms Linkages and Mechanical Controls	Chironis—McGraw-Hill
Racers Encyclopedia of Metals, Fibers, and Materials	Forbes Aird
How to Run a Lathe	South Bend Lathe
Moving Heavy Things	Jan Adkins
Sheet Metal Pattern Drafting	*Daugherty*

Internet Resources

http://www.mcmaster.com/
Hardware, materials, tools. If you can't find what you need here, you have bigger problems.

http://www.metalmeet.com/forum/
Metalworking forum related to cars

http://groups.google.com/group/alt.machines.cnc/topics?hl=en&lnk=sg
CNC machinery newsgroup

http://groups.google.com/group/rec.crafts.metalworking/topics?hl=en&lnk=sg
Active metalworking newsgroup

http://www.practicalmachinist.com/vb/index.php
Active professional metalworking forum

http://www.boedeker.com/
Detailed plastics information

http://www.onlinemetals.com/
Small quantity metals supplier

http://www.freemansupply.com/
Casting and pattern making supplies

http://store.talongripsystems.com/
Work holding

http://www.rjproductsllc.com/rj_vise_soft_jaws.htm
Soft jaw supplier

INDEX